CAMPUS COMPANIES - UK AND IRELAND

Campus Companies - UK and Ireland

DESMOND M. BLAIR
Former Director of Industrial and Commercial Liaison
The Queen's University of Belfast

DAVID M. W. N. HITCHENS
Professor of Applied Economics
Department of Economics
The Queen's University of Belfast

Ashgate

Aldershot · Brookfield USA · Singapore · Sydney

Published by
Ashgate Publishing Ltd
Gower House
Croft Road
Aldershot
Hants GU11 3HR
England

Ashgate Publishing Company
Old Post Road
Brookfield
Vermont 05036
USA

British Library Cataloguing in Publication Data
Blair, D. M.
   Campus companies - UK and Ireland
   1. Industry and education - Great Britain 2. Industry and
   education - Great Britain - Case studies
   I. Title II. Hitchens, D. M. W. N. (David M. W. N.)
   371.1 '95' 0941

Library of Congress Catalog Card Number: 97-78427

ISBN 1 84014 198 0

Printed in Great Britain by The Ipswich Book Company, Suffolk.

Contents

List of Figures

List of Tables

Foreword

In this book the authors have sought to meld the practical experiences of universities and their academic staff in exploiting research through new company formation with the more theoretical views and published outcomes of research in the general field of technology transfer. The study contains a wealth of empirical detail on entrepreneurship, company performance and the management of campus companies, across universities in Britain and Ireland. I believe this will be of great relevance and interest not only to researchers but to those decision makers and managers within universities whose task it is to exploit research through the formation of university companies.

The study examines the development of university companies and presents a taxonomy of company/university relationships which has the potential for practical use to university decision makers in deciding the appropriate support mechanisms required. I have no doubt that the book will contribute to a greater understanding of the motivation of researchers, and their transition from academic to entrepreneur. It examines the university's attitudes and policies relating to company formation and teases out areas where conflicts of interest arise. It then profiles different possible managerial approaches appropriate for different kinds of spin-out companies. Set in the context of survey and case-study material and the experiences of different universities, these approaches are credible.

I have followed the research which led to this book from a number of perspectives. As vice-chancellor of a UK university, during a period of rapid change when the entire funding mechanism of the universities was being revised and important decisions on how to maximise university funding were being made, I can empathise with many of the verbatim statements of university administrators recorded here. As chairman of a university holding company I can well appreciate the frustrations expressed by those who are attempting to inculcate a commercial culture in a traditional university environment and within an academic administrative structure while maintaining and delivering high quality academic research. In the regional and national contexts, I am also acutely aware of the responsibilities universities have to contribute to economic growth. The authors provide a valuable contribution to a better understanding of this complex area.

Sir Gordon Beveridge
Vice-Chancellor
The Queen's University of Belfast
August 1997

Acknowledgements

The authors would like to thank Mary Trainor (Department of Economics at Queen's University Belfast) for editorial help and preparation of the camera-ready copy and Valerie Polding of Ashgate who provided helpful advice.

List of Abbreviations

The following abbreviations and acronyms have been used freely throughout the text. Normally each has been defined at the first instance of use.

AILO	Academic Industry Links Organisation
AURIL	Association of Research and Industrial Liaison
BERD	Business Enterprise R&D
BTG	British Technology Group
CBI	Confederation of British Industries
CEC	Commission of the European Communities (same as EC)
CIHE	Council for Industry and Higher Education
CISAT	Collaboration in Science and Technology
CSO	Central Statistical Office
CVCP	Committee of Vice-Chancellors and Principals
DFP	Department of Finance and Personnel (NI)
DTI	Department of Trade and Industry
EC	European Commission (same as CEC)
EIRMA	The European Industrial Research Management Association
EPSRC	Engineering and Physical Resources Research Council
ESRC	Economic and Social Research Council
EU	European Union
GERD	Total expenditure on R&D in the national territory
GOVERD	Government Expenditure on R&D
HEI	Higher Educational Institution
HT	High Technology
IDB	Industrial Development Board for Northern Ireland
IPR	Intellectual property rights
IRDAC	Industrial Research and Development Committee of the EC

LEDU	Local Enterprise Development Unit (NI)
LT	Low Technology
£b	Billions of pounds, e.g. £1b = £1,000,000,000
£k	Thousands of pounds, e.g. £1k = £1,000
£m	Millions of pounds, e.g. £1m = £1,000,000
NBST	National Board for Science and Technology, Ireland
NEDC	National Economic Development Council
NI	Northern Ireland
NIEC	Northern Ireland Economic Council
NIERC	Northern Ireland Economic Research Centre
NTBF	New Technology-Based Firm
OECD	Organisation for Economic Co-operation and Development
ORD	Other R&D, mainly R&D performed by charities
OSR	Other Services Rendered
QSEs	Qualified Scientists and Engineers
QUB	The Queen's University of Belfast
QUBIS	Queen's University Business and Industrial Services
R&D	Research and Development
ROI	Republic of Ireland
ROPA	Realising our Potential Awards
S&T	Science and Technology
SMART	Small Firms Merit Awards for Research and Technology
SMEs	Small and Medium-sized Enterprises
SMFs	Small and Medium-sized Firms
UDIL	University Directors of Industrial Liaison
UK	United Kingdom
UNICO	University Companies Association

1 Introduction

This book addresses the commercial exploitation of university research through new company formation. It is the outcome of six years research and 13 years practical experience of transferring technology to industry from the Queen's University of Belfast. The research, which draws on the recorded experiences of 25 universities in Europe and America including 18 in Britain and Ireland, was set within the more general context of university/industry collaboration. New company formation was regarded, not as an alternative to more conventional means of transferring technology, but simply as part of a university's repertoire of delivery mechanisms.

Technology transfer influences economic development both nationally and regionally and that is our starting point. In Northern Ireland (NI) for example, where much of the research took place, the local universities are the major sources of research, innovation and new technology and as a result they assume a greater relative importance, as engines for growth of the weak regional economy, than their counterparts in most other regions of the United Kingdom (UK). Although not the first UK university to form spin-out companies[1], Queen's University was in the van of those which set out to develop a portfolio of companies as a deliberate means of commercialising its research. This venture into new company formation began in 1984. Looking, then, for models elsewhere one found a great diversity of types of company, organisational arrangements, methods of funding, and extent of institutional involvement. Some university firms were undoubtedly successful, others less so and yet there was often little more than anecdotal evidence about the factors which influenced their performance. There was little guidance, more than a decade ago, about the issues to be addressed in setting up and running such companies or on the conditions which made for success or failure. Since then university companies, with varying degrees of institutional involvement, have multiplied and become an established mechanism for exploitation of university research.

Many lessons have been learned through practical experience about the processes involved in bringing the results of research to the

1

marketplace, about factors which improve the prospects of success for new start-ups and, in hindsight, a great many things which should have been done differently. Although occasional descriptive articles have appeared from time to time, it is regrettable that few practitioners have documented their experiences in such a way that they can be built upon by others. Researchers by contrast have written a great deal about academic entrepreneurs, the process of innovation, technology as a stimulant to regional economic growth and the role of university research and technology-based firms.

Aims and Objectives of the Research

The aims of the research were:

- To assess the extent to which critical factors identified from the literature as influencing the formation, organisation and operation of university-based companies were relevant to current practice;
- To test the hypothesis that university-linked start-up firms share many of the characteristics of new technology-based firms (NTBFs) but that there are also some unique differences;
- To compare the characteristics of university spin-outs with non-university technology-based companies.

Objectives

The objectives were to add to existing knowledge about university companies, their formation and management by:

- Examining the factors which influenced the universities in establishing spin-out companies;
- Exploring the motivations of actual and would-be academic entrepreneurs;
- Identifying the tasks which had to be carried out before companies were formed and how these were approached by various universities;
- Comparing the processes used and the mechanisms which various universities had put in place to carry out necessary tasks or address critical issues;

- Assessing whether best practice could be identified;
- Identifying similarities and differences between university and non-university based technology-based firms;
- Determining the extent to which technology-based university firms share characteristics with NTBFs;
- Identifying policy implications for universities, for regional development in Northern Ireland and for government nationally.

Methodology

Following a literature review on innovation, economic development, university/industry relations and NTBFs, critical factors affecting the establishment and operation of university companies were identified. A series of structured interviews and face-to-face meetings was then undertaken with university administrators, academic entrepreneurs and company managers to assess the extent to which these factors were relevant to current practice. Included were the motivations and aspirations of the universities and academic entrepreneurs, the mechanisms and organisational arrangements for forming enterprises, the extent of university involvement and the support provided by the institutions to their embryonic firms. This provided insights into the processes, identified the questions to be settled, the issues to be addressed and the elements required in an effective technology-transfer system.

It soon became apparent that there was such a diversity both of types of spin-out firms and university linkages, that to draw any generally applicable conclusions it would be necessary to categorise or characterise them in some way to reduce the number of variables. A method of obtaining a profile for a university/company combination, based on three significant parameters, has therefore been devised and tested. This enables the relative importance to be assessed, case by case, of putting in place infrastructure and support mechanisms.

Two case-studies were carried out. The first, a detailed case-study on the experiences of new company formation by the Queen's University of Belfast and of the activity of its holding company, QUBIS Limited. The second, to provide comparison and contrast, was of campus companies of Trinity College Dublin (TCD).

Following a detailed review of the literature on NTBFs a number of hypotheses were made about the similarities and differences to be expected between university and non-university based high-technology firms. To test these hypotheses a second series of interviews explored the performance of a sample of university firms and the attitudes of the managers and academic entrepreneurs involved to such matters as innovation, employment, competition and access to university resources. Data was obtained on employment, R&D expenditure, sales, added value and productivity. Indicators of export orientation, innovation and new product development were assessed.

These results were then compared and contrasted with similar data from samples of both low-technology and high-technology-based non-university firms located in Northern Ireland. As a further test - and to minimise the effects of the differences in age, sectoral and geographic distribution of the firms in the samples - comparisons were made where appropriate with other published sources of data. The results were then compared against the hypotheses made earlier.

Of particular interest in the context of Northern Ireland is the impact of university-linked firms on local economic development. Although each region has its own unique characteristics, Northern Ireland shares many of the characteristics of other economically weak areas in the UK and elsewhere. Many of the findings should be transferable to other regions.

Arrangement of Chapters

Chapter 2 is a review of relevant background literature on the impact of innovation, research and technology on economic development. Developments in the understanding of the innovation process itself from the simplistic early 'linear' models are described. The recognition of the complexity of the process and the increasing realisation amongst economists, academic researchers, governments and industrialists that innovation involves more that simply technological inputs are significant developments.

Chapter 3 assesses the importance of NTBFs with reference to research findings in Europe and the USA. Their propensity to innovate is noted as being of great significance and also their tendency to cluster. The potential influence of NTBFs in regional economic development is also considered. In Chapter 4 the level of Research and Development (R&D) as an indicator of innovation is discussed and the importance of university research is put into context for the UK.

Chapter 5 is an historical review of developments in university/industry collaboration, mainly in the UK with comparisons where appropriate with events elsewhere. Current issues affecting university/industry relationships are addressed and in particular some of the tensions caused by contradictory signals from government; on the one hand about the role of the universities, and on the other, the way they are funded. A brief summary of the means by which universities transfer technology to industry is included. The characteristics and growth of university-based companies are reviewed in Chapter 6 and developments in the UK are compared with those overseas. A broad taxonomy of university companies is discussed with reference to the literature.

Chapter 7 is essentially a review of literature relevant to academic entrepreneurship and of the characteristics of university-based firms. Critical issues are identified related to the reasons why universities and academics form companies, the motivations of institutions on the one hand and of academic entrepreneurs on the other. Difficulties of transition from academe to the commercial world are also addressed as are the possibility of conflicts of interest for those involved in both spheres. The complementary roles of entrepreneurs, product champions and managers are also considered with reference to the literature.

The results of the first series of interviews are summarised in Chapter 8. The Chapter is entitled 'Towards a Better Understanding of University Companies' and it includes fairly extensive verbatim quotations from those interviewed in an attempt to convey something of the complexity of the issues as well as the range of views expressed. Arising from the interviews it has been possible to identify three distinguishing characteristics which enable a 'profile' for a particular university/company combination to be constructed. These are the degree of university control, the extent of university staff involvement and the nature of the products or services provided by the company. This new tool for 'profiling' spin-out companies and their university linkages is described in Chapter 9. In Chapter 10 the profiling method is used to identify the relative importance of a number of questions which have to be addressed in any university/company situation.

Attention is drawn in Chapter 11 to the holding company model which is emerging as one of the most successful means of systematically exploiting technology through spin-out activity in the UK. A systemic view is used to visualise this model, the various support mechanisms or sub-systems required in particular circumstances, the impact of the wider university system and of the external environment.

Chapter 12 is a detailed case-study of 14 companies which have emerged from the Queen's University of Belfast. The companies are profiled and the group performance is described in terms of growth of employment, turnover and profitability and their impact on the local economy is discussed. The most significant issues which had to be addressed are described together with the impact of the wider university system. For comparison and contrast with QUBIS Limited a similar, but less detailed, case-study of campus companies at TCD was carried out.

Chapter 13 contains the hypotheses made about the characteristics to be expected of university firms and describes the results of the second survey which was carried out to test them. This series of visits was with a sample of 31 university-linked firms and inquired about their attitudes to innovation and new product development, expenditure on R&D, employment, use of university resources, added value and productivity and the impact on their local economies. The results of an outreach programme to some 376 Small and Medium Sized Enterprises (SMEs) in Northern Ireland are analysed in Chapter 14. These companies have been categorised to identify a subset within the sample of 40 technology-based firms. Chapter 15 compares and contrasts the results of the university and non-university companies and draws conclusions for comparison with the hypotheses made in Chapter 13.

Chapter 16 draws the various strands of the research together and contains the overall conclusions which can be drawn from it. This Chapter also contains a discussion of the policy implications for universities, for local economic development and for government generally. Finally some additional areas of research are indicated.

Note

1. The terms 'spin-off' and 'spin-out' are widely used throughout the literature to describe companies emanating from universities. They have no generally accepted definition and are often used interchangeably. The term spin-off has

also been used to describe so called 'spin-off benefits', from space research for example. It is also more appropriately used to describe new firms emanating from established companies. Accordingly the term spin-out is preferred to describe new firms having their origins in universities and is used in that sense throughout this book, except in direct quotations.

2 Technology & Innovation in Economic Development

Introduction

Economists, industrialists, governments and academic researchers agree that early access to new technology and innovation are linked to economic growth both nationally and regionally. The links between a country's level of technological sophistication and its economic performance have been demonstrated in numerous studies (Rothwell and Zegveld, 1981; ABRC, 1987; Segal Quince and Wicksteed, 1988; NEDC, 1989; Roberts, 1991; DTI, 1993; Stoneman, 1992; NIEC, 1993; Kay, 1992; Lipsey, 1993; HMG, 1993; Archibugi et al., 1995; Pianta, 1995).

This is not a new discovery, but the accelerating rate of technological change coupled with the threatened eclipse of hitherto competitive western nations by the burgeoning economies in the far East and Pacific-rim has concentrated minds as never before.

Writing at the start of the 1980s, Rothwell and Zegveld (1981) noted that

> ... there exists a great deal of convincing evidence to suggest that technical change is a prime factor - albeit not the only one - in determining the export competitiveness of a wide range of manufactured goods. A growing realisation of this fact is causing governments in all the advanced market economies increasingly to become involved in instigating measures to stimulate, and assist with, technological innovation in manufacturing industry.

Despite some recent improvement, there is concern both in government and industry about Britain's performance vis-à-vis competitor nations. Citing a study by Solow (1957) which estimated that 80 per cent of productivity growth in the United States during the first half of the 20th century was due to technological innovation, Stoneman (ibid.), claims that, by this yardstick, Britain has had a technological failure in recent years.

The UK's productivity, in terms of real added value per person working in 1980, was lower than all its major competitors.

A commonly used indicator of technological innovativeness is the number of patents filed. The UK's share of patents declined steadily between 1963 and 1988 while that of France, West Germany and Japan all increased. The 1993 UK R&D Scoreboard reported the number of patents issued in USA, to other than American patentees. Between 1980 and 1991 the UK's share of US patents issued dropped from 9.8 per cent to 6.2 per cent while Japan's increased from 24.1 per cent to 46.3 per cent. The percentage increase in the numbers of patents issued during the same period was UK, 15.9 per cent; France, 45 per cent; West Germany, 32.5 per cent; Japan, 194 per cent and the rest of the world excluding USA, 51.3 per cent (Company Reporting Limited, 1993).

In 1993, a joint DTI/CBI report noted the results of a recent survey which concluded that only one in ten British companies was truly innovative. The report lamented Britain's innovatory performance and an assessment that the UK ranked thirteenth out of 22 OECD countries in world competitiveness (DTI, ibid.). The 1994 World Competitiveness Report ranked the UK twelfth for its science and technology base, due to the expenditure and numbers of people employed in R&D but Britain's weaknesses were seen as a lack of qualified engineers and the perception that science is inadequately taught in secondary schools. More importantly the UK was bottom of the table, in that engineering sciences as a profession do not attract young people sufficiently (see also CBI, 1994).

Nevertheless, there is evidence that those companies which do invest in innovation earn a satisfactory return. More than four-fifths of the companies in the 1992 CBI/Natwest Innovation Trends Survey reported market share gains and 78 per cent reported opening up new markets as a result of their investment in innovation (CBI/ Natwest, 1992). It has to be noted also that although employment in high technology-based industry[1] is a small part of overall manufacturing employment, only in Britain and Germany does high technology employment exceed six per cent of total employment in the business sector (OECD, 1994a).

The Innovation Process

Despite the widespread attention which is paid to R&D, innovation and invention are not the same thing and yet the terms are frequently used as if they were interchangeable. Invention is often an important ingredient of, and frequently the starting point for innovation but it is not sufficient by itself. Kenward suggests that the confusion may arise because it is easier to see the process in action when there are physical objects and processes to observe. Inventiveness can be measured but how can a company's innovation be assessed? How, asks Kenward, do you count marketing 'inventions'? The concept of the franchise must be one of the most far-reaching innovations in marketing, but would you call it an invention? (Kenward, 1993).

The OECD Frascati manual defines innovation as

> Scientific and technological innovation may be considered as the transformation of an idea into a new or improved saleable product or operational process in industry and commerce or into a new approach to a social service. It thus consists of all those scientific, technological, commercial and financial steps necessary for the successful development and marketing of new or improved manufactured products, the commercial use of new or improved processes and equipment or the introduction of a new approach to a social service (OECD, 1992a).

Distinctions have been made between radical (or revolutionary) innovation and incremental innovation. Unlike incremental innovation, which is typical of the continual improvements to products or processes which occur in any organisation, radical innovation results from individual inventions and often creates an obvious discontinuity which may be difficult for the organisation to deal with (Fairtlough, 1994). It is important however not to underestimate the importance of incremental innovations. The discovery in 1947 of the transistor was a radical innovation which made possible the modern electronics industry, whereas the subsequent rapid improvements in semiconductor design and silicon fabrication technology - which have driven up performance, reduced costs and led to the burgeoning growth of the industry - would by this definition be classified as incremental innovations. Few would dispute the effect of these incremental innovations yet a further radical innovation may not be made until silicon-based technology is replaced - perhaps by biological-computers.

Two further distinctions - modular innovation and architectural innovation - have been made by Henderson and Clark (1990). Modular innovation represents a substantial step-change of a component within a product or system and architectural innovation occurs when the components are relatively unchanged but their technical inter-relationship shows greatly enhanced sophistication. Extending the previous example, a smaller more compact integrated circuit which occupied a smaller chip would be regarded as a modular improvement to computer technology if the chip was used in the traditional way. If, on the other hand, the innovation which led to the increased miniaturisation could be used to integrate more functions on to a single chip then it would be regarded as an architectural innovation.

However they are categorised, there is now widespread recognition that successful innovation depends not just on science and technology but on a range of other inputs and a complex web of interactions between organisations and people from a variety of disciplines. Chairman of the Sony Corporation, Akio Morita, addressed the confusion between Science, Technology and Innovation succinctly when he chose as his title for the first UK Innovation lecture, *'S' does not equal 'T' and 'T' does not equal 'I'*. He went on to differentiate the separate but often complementary contributions to the process of scientists, engineers and technologists. Innovation, Morita asserted, required creativity in technology, creativity in product planning and creativity in marketing (Morita, 1992).

Roberts and Fusfeld (1981) conclude that five types of role are necessary in innovative organisations - idea generators, entrepreneurs or champions, project leaders, gatekeepers (whose role is to deal with the world outside the organisation and who possess networking and boundary expanding capabilities) and coach sponsors, who are usually senior and experienced people. Some individuals can perform more than one of these roles. The authors argue that innovating organisations must be able to attract, motivate and retain people with these skills.

It is now widely accepted that the traditional 'linear' technology-push or market-pull models - in which either the market simply waits to respond to new products emerging from developing technology or in which researchers produce new products in response to well-articulated market needs - are simplistic. Retrospective studies of innovation have been inconclusive in establishing the extent to which innovations result from science or of the relative influences of technology-push and market-pull considerations. Project Hindsight (based on weapons systems and sponsored by the US Department of Defence) was an early attempt to assess the relative importance of various types of R&D on technical innovation. The results showed that of 700 events only eight per cent were science-based and 92 per cent were designated as technology-based events (Isenson, 1967).

This result is at odds with the outcome of the subsequent National Science Foundation (NSF) sponsored TRACES (Technology in Retrospect and Critical Elements in Science) evaluation by the University of Illinois. This traced the discrete events which resulted in five innovations of unambiguous economic importance i.e. magnetic ferrites, the video tape recorder, the oral contraceptive pill, the electron microscope and matrix isolation. The results showed 70 per cent of 340 events to be non-mission oriented research, 20 per cent mission oriented research and ten per cent development and application work. TRACES would support the technology-push approach (Illinois Institute of Technology Research, 1969).[2]

A study by Battelle Labs in 1973 - a more sophisticated version of TRACES concentrating on 'decisive events' - showed 15 per cent non-mission events, 45 per cent mission oriented and 39 per cent development work, i.e. closer to Hindsight than to TRACES. It was noted that decisive events most frequently occurred at the convergence of separate streams of research activity. When all events were taken into account Battelle fell more nearly between Hindsight and TRACES i.e. non-mission events 34 per cent, mission events 38 per cent and development 26 per cent (Battelle, 1973).

There have been many attempts to explain the ambiguity of these findings. One explanation may lie in the different time frames in which the two analyses were made. TRACES indicates that a time span of perhaps 30 years or more is needed fully to assess the impact of science on technology. Stankiewicz (1986) suggests that this may be interpreted as either that the transfer of knowledge from science is intrinsically very slow, or that it tends to be greatly delayed due to poor interaction between universities and industry.

The long timescale often associated with basic research and its importance in providing the underlying knowledge base to enable technological advances to be made is stressed by Pavitt (1992). He argues that basic research plays an important role not by contributing directly to the finished product but by providing the knowledge, tools and techniques to solve problems, backed up by networks of trained researchers. Pavitt argues that we cannot rely on market demand to lead to the necessary technological advances and sees it as government's role to underpin basic research.

There is also frequently a blurring of the distinctions between and relationships of science and technology. Science and technology are different but inter-related and overlapping systems (Pavitt, 1990). It is not sufficient to regard technology simply as the ability to carry out productive transformation (Metcalfe, 1995). The widespread view that technology is simply applied science or that it can always be explained in terms of science has also been challenged by Gibbons (1984) who offers the proposition that technology is autonomous, that it is a form of knowledge with a life of its own.

> Thus it becomes possible to speak of technological discoveries - that is, of new insights into ways of making or doing things which derive their inspiration from the field of possibilities presented by the technology itself ... The mere fact that technological knowledge is not usually systematised and presented as a body of theoretical knowledge is certainly not sufficient grounds for regarding it as of secondary importance to science (Gibbons, 1984).

Gibbons sees no evidence that science is becoming more important as a source of new ideas for technological innovations. In some areas, such as the mechanical and electronic industries, it seems that technological inventiveness is the prime source of new products. There is also the historical observation that many of the innovations of the industrial revolution were more the work of 'inspired tinkerers' than of scientists. A model is proposed in which two streams flow through time, one of science and one of technology They can proceed independently but may also interact or assist one another. There is some evidence that, in the two stream model, during the early stages of development the propensity to interact is high and there is a great deal of cross fertilisation (Gibbons, op. cit.).

The innovation process has been the subject of a number of reports and publications by government, aimed at industry and generally of an exhortationary nature. These have tended to cite examples of successful innovation, frequently by companies overseas, and have sought to show that R&D activity is linked to profitability and growth. The importance of good communications between industry and investors, the need to take a long term view of new product development and the desirability of having appropriate attitudes to innovation built into company and the national culture are all emphasised. Checklists and examples of successful practice abound in these and similar publications (McKinsey, 1991; DTI, 1993; IABATC, 1993).

Notwithstanding these examples, it is not clear that processes that have worked at particular times in one set of circumstances can be extrapolated appropriately to others. The reality is that there is no consensus about what constitutes 'best practice' although various interactive models with feedback loops and iterative processes have emerged, see for example McKinsey (ibid.); Kline and Rosenberg (1986) and Goddard et al. (1994).

Five generations of the innovation process have been charted by Rothwell (1993) who argues that - although this does not imply a sequential process and all five generations of innovation process co-exist in various forms - those companies which master the fifth generation process will become the leading innovators of tomorrow.

Table 2.1   Five generations of the innovation process

Generation / Process	Period	Characteristics
1 Technology push	1950s to second half of 1960s	Simple linear process. Emphasis on R&D. Market is a receptacle for the fruits of R&D.
2 Need-pull	Second half of 1960s to first half of 1970s	Linear sequential process. Marketing emphasis. Market is source of ideas for directing R&D. R&D has a reactive role.
3 Coupling model	Mid-1970s to mid-1980s	Sequential but with feedback loops. Push or pull or push/pull combinations. R&D and marketing more in balance. Emphasis on integration at the R&D/Marketing interface.
4 Integrated model	1980s to early 1990s	Parallel development with integrated development teams. Strong upstream supplier linkages. Close coupling with leading edge customers. Emphasis on integration between R&D and manufacturing (design for makeability). Horizontal collaboration (joint ventures etc).

Table 2.1 cont...

Generation / Process	Period	Characteristics
5 Systems integration and networking model (SIN)	Early 1990s - ?	Fully integrated parallel development. Use of expert systems and simulation modelling in R&D. Strong linkages with leading edge customers (customer focus at the forefront of strategy). Strategic integration with primary suppliers including co-development of new products and linked CAD systems. Horizontal linkages: joint ventures; collaborative research groupings; collaborative marketing arrangements. Emphasis on corporate flexibility and speed of development (time-based strategy). Increased focus on quality and other non-price factors.

Source: Rothwell (1993)

Rothwell asserts that neither large nor small firms are inherently better at innovation, there are advantages and disadvantages for each. In large firms innovatory advantage is associated with greater financial and technical resources, in small firms with flexibility, dynamism and responsiveness. In other words the advantages of large firms are mainly material while those of smaller firms are mainly behavioural. Large firms which can combine material and behavioural advantages are in a very strong position in terms of techno/market dynamism. Large/small firm combinations can assist in overcoming disadvantages and provide complementary benefits.

A weakness with all the interactive models, pointed out by Scase (1992) is that, although laudable in theory they are very general in practice and offer little guidance about how to develop the recommended skills and forms of organisation which the models prescribe. The importance of integrating R&D into the core activity of the firm as a strategic entity rather than a support unit is stressed by Scase. The need to understand the interpersonal dynamics is also important.

> Clearly innovation involves both internal and external organisational processes and there can be little understanding of it as an organisational process without consideration of, for example, interpersonal networks, the social, economic and political relations within firms as well as a wide variety of hidden beliefs and assumptions embedded in organisational cultures... More qualitative research is clearly needed into organisational processes if we are to understand the interactive paradigm and its value (Scase, 1992).

Fairtlough (1994) in a wide-ranging literature review of organisation for innovation concludes that success depends on effective interconnections between many groups of people within and without the organisation and that the priority for these connections varies with the type of innovation being attempted. A 'loose-tight' rule is recommended. Organisation structure should be loose i.e. decentralised, but tight in the sense that once a consensus about priorities has been reached, everyone should respect them. These issues are particularly important in the organisation and support systems required for university spin-out companies and will be addressed later in that context.

Implications for Policy

So what options are available? Porter (1990) stresses that to improve national competitiveness the emphasis should not simply be on science or technology but that it is necessary to create a policy for innovation. He lists seven characteristics of such a policy:

- A match between science and technology policy and the patterns of competitive advantage in the nation's industry;
- Emphasis on research universities instead of government laboratories;
- Principal emphasis on commercially relevant technologies;
- Strong links between research institutions and industry;
- Encouragement of research activity within firms;
- Primary emphasis on speeding the rate of innovation rather than slowing diffusion, for example, by avoiding undue emphasis on protection of intellectual property;

- A limited role for cooperative research.

For the UK specifically, Porter recommends that R&D funds should be channelled through universities and specialised research institutes and not into direct subsidies. He also argues that a reallocation toward commercial R&D will be necessary to restart the upgrading of competitiveness.

Supporting a more micro-based approach, Stoneman (ibid.) argues that Britain's failure to develop a strong technological base has been at least partly the result of a concentration, by successive governments, on short-term macro-economic demand policies. These Stoneman describes as anathema to technological investment. He argues for a policy which would concentrate on stimulating civil R&D in industry, direct funding or risk-sharing within a coherent strategy that is consistent with existing industrial and training policies. A diffusion policy should also be considered in areas where there are 'technological gaps'.

Such a risk-sharing approach could well run into difficulties arising from conflicts of interest between national policies, on the one hand and on the other, what individual firms perceive as their own economic interest. For example, Lipsey (1993) has drawn attention to the changes which occur as one technoeconomic paradigm[3] begins to be replaced by others.

Not least of these is the fact that the first investments in the new paradigm are more risky and yield lower returns that those which follow. This, combined with the 'endogenous' nature of R&D - which responds to prices and opportunities for profits to be made - means that early decisions are not only risky but can have major long term effects. Technology which has attracted funding and been substantially developed may not be abandoned even when a superior technology subsequently comes along. There is no way of knowing in advance which technologies are going to be successful. The technology which gains an advantage at the early stages may attract enough R&D to forge ahead even if some other technology might have proved superior given equal R&D support. Since there is a lack of knowledge about technologies until they are tried, Lipsey argues that open competition in the market does not guarantee the best result and there is potential for market intervention. See also Rosenberg (1982) for a development of the factors which determine and influence technological change. The observation by Arthur (1990) that those parts of the economy which are knowledge-based are largely subject to increasing, as opposed to diminishing returns, is also relevant in this context. Arthur quotes the example of the early competition in video recorders between VHS and BETA technology and argues that even if, as was claimed, the BETA technology was superior the initial gains obtained by the VHS system enabled that technology to succeed. The main point of the argument being that under conditions of positive feedback instability occurs and there is no way of predicting which of several possible equilibrium conditions will be reached.

A major plank of the UK Government's White Paper, 'Realising Our Potential', was a programme of Technology Foresight to be conducted jointly by industry and the science and engineering communities. The intention was that the results of these deliberations would inform government's decisions and priorities. The hope being that the exercise would forge a new working partnership between industry and the scientific community as well as providing early notice of emerging key technologies (HMG, 1993). The initiative was, by and large, welcomed publicly by the parties to the proposed dialogue although privately there was some scepticism about what could be achieved (Kenward, 1995). It was interpreted by some as a process of 'picking winners'. There is a good deal of support for the view that winners select themselves and success depends as much on timing, innovatory marketing and distribution, pragmatism and luck as on orchestrated togetherness between industrialists and scientists. That, however, is a rather cynical view, although not without anecdotal evidence to support it. A more rational, or at least a more academically respectable, concern about the process is that it is by no means established that all, or even most, commercially viable innovations arise from science (Gibbons, ibid.; Irvine and Martin, 1984). At a practical level the concerns of Lipsey (op. cit.) about the reluctance of companies to abandon technology in which they have invested heavily may also be relevant to the outcome of the Foresight exercise.

Supporting technology foresight, Newby (1993), has drawn attention to the complexity of the process of innovation in the first of four considerations, arising from ESRC research, which should influence policymakers, i.e.

• The interactions between science, technology and economic competitiveness are more complex than is often realised. More detailed research and analyses of how innovation works in practice is called for;

- The need to put science in a social and economic context. Science does not operate in a vacuum but is influenced by a host of social and economic factors. Their effects on innovation need to be identified. It is also important to recognise that innovation can be driven by markets. The 'divergent' processes of the scientific community need to be married to the 'convergent' processes of innovative industrial development;
- The importance of human resources must be taken into account. Innovation does not happen on its own. It has to be managed and organised in a flexible and coherent way where everyone shares a sense of purpose.
- Skills shortages have contributed to Britain's lack of competitiveness. Research shows that poor training, especially at the lower end of the skills spectrum has been a problem. Other studies show that companies employing a higher proportion of graduates and those with post-secondary education performed better.

Newby refutes the notion that Foresight is about identifying winners and views it as a way of scanning and systematically analysing generic technologies which firms could consider in their research priorities. He also asserts that knowledge transfer, rather than technology-transfer is a key component in the process of innovation. On the other hand the UK is not the first nation to engage in this type of Delphic process. As science and technology as well as markets become increasingly international, one would have to ask whether the emerging technologies determined as being of significance to Britain could be markedly different from those which matter to other industrialised countries.

The results of the Foresight panels' deliberations are now available in fifteen volumes which, although variable in quality and prescience, attempt to set out where technology and markets are leading in the various industrial and economic sectors during the next decade and beyond (OST, 1995). Attention to Foresight priorities is now becoming a requirement for many UK Government supported R&D schemes. In 1994 the UK Research Councils introduced Realising our Potential Awards (ROPA) for R&D to universities which could demonstrate that they had carried out research for industry valued at least £20k in the previous year. Although the £46.7m awarded in the first round was taken up in a rather patchy manner by the universities, additional funds were to be added to make £70m available in the next round (Research Fortnight, 1995).

There is a concern in academia that the almost inevitable market-driven - or at best strategic - focus of many of the sectoral Foresight reports will result in a diminution of the perceived importance of, and therefore of funding for, more basic research. It should not be lost sight of that many of the really seminal scientific breakthroughs, such as the understanding of DNA, have come from purely curiosity-driven research. Pavitt (1992) believes that basic research is essential for international competitiveness and that there is a strong case for supporting curiosity-driven research which has no apparent commercial application. He stresses that, although world class basic research is necessary if we are to have world class technology, policies are needed to encourage firms to cultivate world class technological capacities, resources and strategies to translate the research into commercial realities.

Pavitt advocates that national policies should follow three principles:

- Broad funding decisions on basic research in science and engineering should include assessment of the potential economic and social benefits of research training, techniques and instrumentation, linkages to international excellence, engineering disciplines, business and other practitioners. The policies should take into account the diverse technological origins of each product and the diverse scientific origins of each technology;
- The dominant criteria for allocating resources between institutions, programmes and projects should be research excellence, strong postgraduate training and links - often informal - with industrial practitioners. Priority should not be given to government laboratories and private firms since these institutions are weak in postgraduate training;
- International ties should be fostered to enhance Britain's research base, there should be:

  - greater use of high quality foreign scientists and engineers in the reviews and evaluations of British basic research;
  - greater resources for foreign travel, postgraduate training and sabbaticals, with the specific aim of improving skills;

- greater encouragement of foreign firms performing and commercialising R&D in the UK, particularly where British firms do not meet international standards.

The Chairman and Chief Executive Officer of Olivetti, Carlo De Benedetti, stressed the need for innovation at two levels

> A country that has an established manufacturing industry, and that wishes to continue as one of the leading industrial nations, must have an industrial policy which encourages primary innovation (through basic and applied research, creation of new products, independent development of new production processes and instruments) and secondary innovation (through distribution and adaptation of the economic system). Innovation also means organisational and managerial change (De Benedetti, 1987*)*.

Benedetti's latter point, that successful innovation requires more than just good technology, is one that both industry and governments have been slow to recognise. It is worth re-emphasising that a holistic approach to innovation is required to ensure that the disparate but complementary skills of scientists, engineers, entrepreneurs, financiers, marketers and managers, meld to produce something which is more than simply the sum of their separate contributions.

In a later publication, Newby (1995) summarised four insights into the management of innovation arising from ESRC research i.e. that regeneration is possible; firms' strategies are important but equal emphasis should be placed on building social institutions; managerial expertise is vital to manage the process and cooperation and collaboration are just as important as competition.

Examining the economic rationale for an enhanced Irish National Science and Technology (S&T) capability, Kinsella and McBrierty (1995), conclude that the social as well as the natural sciences play an indispensable role in maximising the benefits of S&T in society.

Rothwell and Zegveld (1981) have also pointed to the dangers of relying on inputs from technologists and scientists alone

> Good high level natural scientists and technologists are clearly essential but they are not enough. Indeed it could be dangerous if they were to be a unique source of advice and policy formation. Competence in natural science and technology must be systematically related to other types of expertise in economics, social policy and politics. The development of this all round

competence is of course equally essential in industry and in both sectors there is clear evidence of the development of management techniques, institutions and structures, appropriate to the nature of these problems.

Although there are differing prescriptions for policy, on the importance of innovation itself there is a rare unanimity between academics, industrialists and government. The CBI/DTI Innovation report (ibid.) declares that

> Innovation is not a new concept and is not synonymous with invention, although invention may be part of the innovation process. It is synonymous with good management of all the company functions - research, design, development, production purchasing, marketing, sales distribution servicing, training, finance and administration. ...Innovating firms tend to have much larger market shares and higher growth rates and profits, and a link has been demonstrated between R&D expenditure and subsequent sales revenues relative to those of competitors.

Technology in Regional Development

Considerable emphasis has been placed on technology as a stimulus to regional economic development and although there is a prima facie case to support such a policy, the results in the UK have at best been mixed and there are those who question its effectiveness. e.g. Ewers and Wettmann, 1980; Oakey and Rothwell, 1986; Harris, 1988; NEDC 1989; DFP, 1989; Keeble, 1993.

Harrison and Harte (1990) identified two separate contexts within which regional policies based on innovation and technology operate:

- 'Entrepreneurial' mode in which new start-of-cycle industries emerge, characterised by novel products and markets which are so novel as to be still in the process of definition;
- 'Managed' innovation in which end-of-cycle industries develop new products based on relatively new technologies for established markets.

The latter context is broadly concerned with ensuring that firms across a wide spectrum of industry keep abreast of new and emerging technology to remain competitive.

Technology Diffusion

It is a moot point whether technology-based economic growth should focus on the use of technology to produce new products or services or on its application to other industries (Segal Quince and Wicksteed, 1988). Amongst the main structural weaknesses in many regions is the lack of innovation capability in smaller firms. Arguing from a German perspective, Herdzina and Nolte '(1995) assert that one of the most important components of an economic policy in a peripheral region should be to provide an adequate infrastructure to encourage and assist technology-transfer and innovation.

There are certainly many examples which demonstrate the effectiveness and the competitive edge which can be given to a company by the transfer and application of technology, which although novel to the recipient's industrial sector, is commonplace in another. The introduction of laser printing to label collar sizes by the shirt manufacturers is one example. The use of 'commonplace' infra-red technology by bacon producers to measure the thickness of pig fat is another (Beveridge, 1991).

The importance of this type of technology diffusion to the many firms which can benefit from applying what, to them, is new technology must not be lost sight of when considering the effects of innovation on regional growth. The availability of novel technology may be as important as the creation of new products and processes to firms, and particularly to SMEs in peripheral regions. Its importance at national level has been recognised in the Government's White Paper on its strategy for the UK Science Base (HMG, 1993 paras, 1.18 [4] and 2.5).

Entrepreneurial Growth

Turning to the 'entrepreneurial' mode of growth, it is unfortunately the case that the external technological environment for companies in the peripheral regions of the UK is much less rich than for those in the more advanced regions (Rothwell, Dodgson and Lowe, 1989).

Harris (1988) compared innovations by region within the UK during the period 1945-83 and concluded that

> The South East region dominated other regions of the UK not only in terms of
> numbers of innovations produced, which may in part reflect its relative size,

but also on the basis of other indices of relative performance, such as specialisation and 'innovativeness'.

Studies for the STRIDE committee of the EC showed that, in 1983, 45.5 per cent of industrial firms' R&D units were in the South East and, in 1984, 51 per cent of professional engineers and scientists in the engineering industries were employed there.

It is not surprising that the South East, which has most of the larger firms and a disproportionate number of company headquarters and central research facilities should produce most innovations. Less obvious perhaps is the cumulative effect of this disparity as existing technology-based firms, which are a prolific source of new entrepreneurs, increasingly beget others and exacerbate the imbalance between the regions. This is bad news for regions less well endowed with a core of high-technology companies to start the process. Because of this uneven distribution, weaker regional economies face great difficulties in their efforts to develop a share of innovations proportionate to their population.   Oakey, Rothwell and Cooper (1988) stress the need, in these circumstances, for development agencies in depressed regions to provide intensive help and encouragement to the meagre level of entrepreneurship which does exist, as well as to take steps to add to this 'natural' level.

Deprived of an industrial seedbed for entrepreneurship, peripheral regions must look elsewhere for a technological stimulus if their position vis-à-vis more prosperous regions is not to deteriorate. One readily available source is the technological spin-out activity from regional universities which may well be the major source of regional research and innovation (Beveridge and Blair, 1994). There is however a caveat, because regions in which high technology firms are concentrated often have higher than average performance in other sectors as well (Monck et al., 1990). Whether this is because the high technology firms set the pace for others or - as seems more likely - other environmental factors are at work, this effect may prove to be a limiting factor on the potential to raise the economic performance of less favoured regions through encouragement of high technology-based entrepreneurship.

A further difficulty is that most product innovation tends to be carried out in central R&D facilities of larger companies, usually at their headquarters. Branch plants in peripheral regions, by contrast, are more likely to concentrate on process innovations aimed at cost reductions, quality, or productivity improvements (Harris, ibid.). The relatively few larger technology-based firms in peripheral regions which engage in state-of-the-art development of new products and processes, can nevertheless create considerable peripheral growth in companies which support them. Not surprisingly, therefore, the larger companies are well looked after by government. Regional development agencies compete to create favourable inward investment incentives to attract more such firms and to provide favourable regimes for growth.

The increasing importance of SMEs in a 'Europe of the regions' in which local conditions will have a major impact on small firms and their ability to be effective as agencies for technology-transfer is noted by de Ridder (1995). He urges that regional policies should aim to improve the social and cultural climate in which entrepreneurs and their staff live, work and communicate.

Summary

There is broad agreement about the importance of access to and application of new technology for economic growth, both nationally and regionally and evidence that companies which invest in innovation earn a satisfactory return.

The innovation process is much more complex than was once believed. Linear models derived from market-pull or technological-push are simplistic and the process is recognised as being an iterative one requiring more than simply technological inputs. The latter must be complemented and supplemented by expertise in the social sciences and the provision of appropriate management and support structures.

There is a good deal of agreement, but no consensus in the literature about the implications for national policy. Recommendations include:

- The need to speed up the rate of innovation, to channel R&D funds through universities and research institutes rather than direct subsidies (Porter, 1990);
- Place less reliance on short-term macroeconomic demand policies and concentrate more on stimulating civil R&D in industry and a diffusion policy to fill technological gaps (Stoneman, 1992);

- Because there is no guarantee that open competition in the market will ensure the best result there is potential for market intervention (Lipsey, 1993);
- World class basic curiosity-driven research must be supported to maintain international competitiveness and the dominant criteria for allocation of resources between institutions, programmes and projects should be research excellence (Pavitt, 1992).

The UK Government has backed a Technology Foresight Programme which aims to build stronger links between industry and the research community aimed at identifying those technologies and applications which will be significant beyond the millennium (HMG, 1993).

Technology is also seen as a stimulant to regional economic growth, on the one hand through the diffusion of new or novel technology to existing companies and on the other the encouragement of new companies and industry through the commercial exploitation of new and emerging technologies. There are however other environmental factors which influence and may limit the effectiveness of this approach to lift regional economic performance.

Because most technologically based firms in the UK are located in the South East of England, peripheral regions are relatively deprived of a prolific source of NTBFs. In these circumstances regional universities become particularly important as seedbeds for NTBFs and engines for economic growth.

Notes

1. Most definitions of high technology companies include criteria such as the proportion of engineers and scientists employed and the R&D intensity - measured as the share of production, added-value or percentage of sales. A disadvantage of these measures is that they cover only manufacturing industry and classify as low technology those firms which, although not engaging in R&D, make use of and may be heavily dependent on technologically sophisticated equipment. A study by the Department of Finance of Canada, (1992), attempted to rectify this by including the proportion of high-technology inputs embodied in final goods and services (see OECD, 1994a, page 97; Butchart, 1987).

2. Comparing the Hindsight and TRACES studies is complicated by the different terminologies used. It is only a slight over-simplification to liken

the 'science-based' events in Hindsight to the 'non-mission oriented' events in TRACES and conversely the 'technology-based' events to the 'mission-oriented' events. Viewed in this way the apparent contradictions between the two conclusions are obvious.

3. A technoeconomic paradigm involves a complex interaction of some key products of wide application, key materials whose costs are falling, a means of organising economic activity, a supporting infrastructure and typical patterns of industrial concentration of efficient location (Lipsey, ibid.; Freeman and Perez, 1988).

3 The Importance of New Technology-Based Firms

Overview

There is ample evidence of the importance of NTBFs to economic development. Their influence has been particularly marked on the United States economy since the early 1950s, notably in computers and information technology and more recently in biotechnology. By contrast, a study in 1977 estimated that there were then only 200 NTBFs employing 15,000 people in the UK and 100 in the FRG employing 12,000 (Little, 1977). However, as only manufacturing firms were included in the study these figures are probably an underestimate. A later study covering the period 1970 to 1985 demonstrated that there had been dramatic growth of NTBFs in the UK since the mid-1970s, most of which had been formed since 1979. There were some 7,000 such companies, employing about 120,000 in the UK in 1985. In Germany the number of firms had risen to 3,000 and in both countries there was significant geographical clustering (Segal Quince and ISI Institute, 1986).

It is believed that one reason for the rise in NTBFs in Europe since the early 1970s is due to the formation of emerging 'technology clusters' such as biotechnology and information technology. Unlike experience in the USA however, employment in high-technology industry has not grown significantly in Europe and in the UK has declined by almost 11 per cent between 1981 and 1991. This is believed to be due to reductions in defence-related expenditure as well as rising productivity (Keeble, 1993).

Characteristics and Performance of NTBFs

Data from the SPRU Innovation Database, which contains information on significant British innovations, has been used to show that the share of 'important innovations' originating from small and medium-sized firms

31

(SMFs) and from NTBFs has increased steadily and significantly and that both categories are playing an increasingly important innovative role in the UK. The most notable feature being that about a third of such innovations come from companies having 1-199 employees. For all SMFs i.e. firms employing 1 to 499, the innovation share increased from 22.6 per cent during 1965-69 to 29.2 per cent during 1975-79, to 38.3 per cent during 1980-83 (Rothwell and Dodgson, ibid.).

Studies from a number of countries have demonstrated an association between innovativeness and job creation, e.g. France (Piatier, 1981), Canada (De Melto et al., 1980) and the Republic of Ireland (NBST, 1980). A study in the Republic of Ireland by the National Board For Science and Technology constructed an 'innovation index' for 120 firms employing less than 50 people. Negative employment growth correlated with lack of innovation while employment in the more innovative firms grew faster. The rate of growth was strongly correlated to innovativeness and the younger firms in the sample were more innovative than those less than 30 years old (NBST, 1980.) There is evidence that high-technology-based firms grow faster than low-technology firms and that amongst high-technology firms 'younger' firms have a better employment generating performance than 'older' firms (Oakey and Rothwell, ibid.; Morse, 1976; American Electronics Association. 1978; Rothwell and Zegveld, 1982).

There is however a fairly high mortality rate amongst NTBFs, more than 30 per cent ceasing to trade after three years (Ganguly, 1985). Rapid growth tends also to be confined to a relatively small proportion of such firms. A variation of the Pareto principle applies in that the fastest growing four out of every 100 small firms will create half of the jobs in the group over a decade (Storey et al., 1987).

Despite the preoccupation with high-technology firms by government and development agencies in recent years, relatively few in Britain have grown significantly to become 'flagship' firms in their sector committed to domestic research, development and production. Oakey (1991a) argues that this failure to grow significant indigenous companies has led to a preoccupation with numbers of firms rather than with their quality, judged in terms of rapid individual growth. Furthermore many of those firms which do possess potential for rapid growth are introspective in their attitude to expansion, preferring to rely on self-generated profits to fund growth. In most cases new firms do not contribute substantially to employment.

A study carried out by SPRU in 1987 for the Industrial Research and Development Committee of the European Community (IRDAC) examined the strategies and performance of 12 leading technology-based, 'niche strategy' companies (Rothwell and Dodgson, 1987). These companies, not all of which were new start-ups, exhibited the following characteristics:

- Even long-established companies can enjoy renewed levels of high performance by adopting appropriate techno-market strategies;
- The survey companies showed remarkable growth in annual turnover between 1983 and 1985;
- Total employment grew in the 12 firms by over 1,000, equal to 30.8 per cent, between 1983 and 1986;
- Generally profit levels were high;
- Generally the companies were highly export-orientated;
- The companies all devoted a considerable percentage of resources to R&D;
- A high number of graduate engineers and scientists were employed in R&D;
- The youngest firms spent the highest proportion of resources on R&D;
- As well as maintaining strong internal R&D, considerable emphasis was placed on accessing external technological information, including feedback from users and suppliers.

Although successful, all the companies had experienced problems in their development, amongst which were:

- Funds for growth or to finance vertical integration had been difficult to obtain;
- The companies that experienced severe financial or market difficulties did so largely because of poor or complacent management;
- The most widely articulated concern about future obstacles to growth related to shortages of key personnel, including lack of entrepreneurial management (six firms), shortages of electronics engineers (five firms) and marketing managers (four firms).

The results of the IRDAC study confirmed that

... even these high-performance companies suffered a number of growth-related problems, two of which related to lack of information on potential collaborators abroad and difficulties in obtaining export market information, especially on the all-important American market. Significantly, all of these firms had been successful in accessing external technological know-how, which was facilitated through their high levels of in-house R&D employment (Rothwell and Dodgson, 1987).

A study, by Mustar (1995), of enterprises established by researchers in France revealed that 50 per cent generated some of their income from exports, two-thirds spent more than 20 per cent of their turnover on R&D and that these companies maintained close links with the academic world as a source of expertise and technological input.

A questionnaire in Sussex showed that links with academic institutions increased positively with the number of qualified scientists and engineers (QSEs) employed (Lowe and Rothwell, 1987). However, to be able to access external information and know-how requires a certain capability and it has been suggested that the ability of smaller firms to do this may be limited by the extent to which they have suitable qualified engineers and scientists.

Oakey et al. (ibid.) point out that the information requirements of SMFs vary enormously depending on their technologies and strategies. They identify three basic operating strategies adopted by SMFs:

- Sub contractors, whose information needs and technology derives largely from their customers;
- Traditional small firms in long established areas, such as textiles, garments, leatherware and metalworking. Most will not carry out formal in-house R&D but will be involved in incremental development activities;
- Technology-based small firms which compete on product based technology. These companies carry out in-house R&D and must be innovative to survive.

The IRDAC study highlighted many of the problems and issues affecting NTBFs, not least their needs for external information. These matters will be referred to later and the results of the IRDAC study will be compared and contrasted with the experiences of high-technology university spin-out companies.

NTBFs in Regional Development

In regions lacking significant numbers of larger technology-based companies from which NTBFs might be expected to spin-out, it is not surprising that attention has turned to the contribution which can be made to the local economy by smaller firms. NTBFs are worthwhile not only as such, but as large firms in prospect. Despite the poor record of the UK in creating flagship companies, it would need only an occasional home-grown 'DEC' or 'Hewlett Packard' to transform many peripheral regions. It is well worthwhile therefore to examine the nature of NTBFs in the context of regional development.

These companies have considerable potential for job creation, a major consideration in depressed regions. The propensity of high-technology firms to cluster has already been noted and for most of those in the UK to occur in the South East of England. The extent to which the effects of this geographical disparity can be counteracted depends largely on the effectiveness of local external environmental factors to stimulate innovation and entrepreneurship.

One hopeful factor is that there is some evidence that NTBFs in less well developed or 'hostile' regions, although fewer in number, once established can actually grow faster than their counterparts in more favoured locations, Vaessen and Wever (1993). It has been speculated that one reason for this is that firms in less well endowed regions face fewer competitors and have specialised in niche markets. Also there has been a marked spatial shift in location of NTBFs from urban to rural and small town environments. Although noting the importance of a supportive infrastructure and research linkages with, inter-alia, universities and other major research organisations, other factors are at work, including the role of attractive residential environments (Keeble, 1993). It is believed that there is quite a high 'multiplier' effect on employment in other sectors. Although new high-technology companies may not themselves employ many people, they may create a faster growing and economically more significant service sector around them.

> Both directly and indirectly they are significantly greater generators of new jobs than 'conventional' small firms, and they have a significantly higher export (and import substitution) potential (Segal Quince Wicksteed, 1988).

A possible inhibiting factor arises from the belief that, because access to information is crucial to innovation, those firms based in larger agglomerations will have an advantage vis-à-vis those in rural or peripheral regions. However it is also argued that as firms become more mature these agglomeration advantages matter less and there is a filter-down effect as firms move to more rural regions. A study in the Netherlands failed to prove that the probability of a firm engaging in R&D was influenced by its location although there was some support for the urban hierarchy and filter-down theory amongst service firms (Kleinknecht and Poot, 1992).

All of these supposed benefits are not without controversy. There are those who question the supposed job creation of technology-based economic development and there may be other explanations for the exceptional growth reported in some studies. Nor is the definition of a high-technology company universal amongst researchers and there is an important distinction to be made between, on the one hand, companies developing new technology-based products and on the other, those employing technologies (Segal Quince Wicksteed, ibid.). Differences in definition of what constitutes the high-technology sector can lead to different conclusions about its size in particular regions and therefore about its potency. One must be careful therefore about drawing conclusions from studies based on different definitions. For definitions of high-technology firms see for example, Butchart (1987) and OECD (1994a).

Nevertheless, although not a panacea, there is little doubt that technology-based economic growth is a real phenomenon. It is however dependent upon and constrained by the wider environmental and industrial milieu. A major objective of regional economic development therefore should be to maximise the growth potential for high-technology companies over what would occur 'naturally' if the free market were left to operate by itself. This can be done by increasing the number of new companies and/or picking winners i.e. identifying and stimulating the growth of particular companies or technologies. In Northern Ireland for example, the Local Enterprise Development Unit (LEDU) has identified, for special support, companies which are believed to have potential to double their turnover in three to four years (NIERC, 1996a).

In either case the local universities have an important role to play and it is apparent that in weak economies, which lack a core of companies regularly engaging in R&D and from which new entrepreneurs can emerge, the universities become the major source of innovation and of potential NTBFs.

Summary

There has been a dramatic increase in NTBFs in Europe during the 1980s and 1990s, one reason for which is believed to be the emergence of technology clusters such as biotechnology and information technology. Disappointingly however, in contrast to experience in the USA, employment in high-technology firms in the UK has actually declined in recent years. Although it has been shown that technology-based firms grow faster than others, few NTBFs in Britain have grown to a really significant size i.e. to the extent that they have a major influence on their sectors.

Small and medium sizes firms (SMFs) tend to be more innovative than larger firms, and there is evidence that employment growth is correlated with innovativeness, although later research suggests that this growth tends to be confined to a minority of such firms. Four out of every hundred firms will create half of the new jobs over a decade. There is also a fairly high mortality rate amongst NTBFs.

Other characteristics of NTBFs demonstrated by various researchers include, satisfactory profitability, high export-orientation, a high proportion of expenditure on R&D, high numbers of graduates employed and a need for external technological information. Typical problems reported include difficulties in obtaining funds for growth, financial difficulties, poor management and shortages of key personnel. Their potential for growth makes technology-based firms an attractive proposition as means of stimulating regional economic growth. This is particularly true in weak economies seeking to build a critical mass of technology-based businesses. Some research suggests that, although fewer in number, NTBFs once established in 'hostile' regions can actually grow faster than their counterparts in more favoured regions. Another attractive feature in the context of developing regions is the high employment multiplier effect in other sectors.

Technology-based firms, especially in their early years of development, require access to appropriate technology and assistance and this must be provided if they are to thrive. Regional economic policies should therefore put in place support mechanisms to ensure that the growth of technology-based industry exceeds that which would occur 'naturally'. The universities have a major role to play as support agencies and sources of technological information and in regions which lack a core of companies which regularly engage in R&D they become the main source of innovation and potential NTBFs.

4 R&D as an Indicator of Innovation

International Comparisons

Although not sufficient by itself, research is necessary to stimulate technological innovation. Porter (ibid.) for example, in a study of competitive advantage of ten leading industrial nations, identified a strong R&D capability as a common and essential characteristic of an effective science and technology policy.

In a study of 20 OECD countries covering the period 1970-90 Pianta (1995) found strong links between R&D intensity, GDP per capita and R&D expenditure per employee. A regenerative 'virtuous circle' is postulated in which research and innovation sustain a country's technological capability, contribute to its accumulation of capital which leads to economic growth. However if the virtuous circle is to be sustained as countries achieve high income and productivity levels and approach the technological frontier, it is likely that a greater role will be played by other factors such as knowledge, learning processes, human capital, quality of research, organisational innovations and favourable institutional conditions.

Gittleman and Wolff (1995) found that R&D activity was significant in explaining inter-country differences amongst the more advanced nations but was of little help in explaining the performance of less developed countries. Furthermore the returns to R&D had fallen off after 1970 amongst the advanced nations. The route to improved productivity growth, it was thought would be from software-related developments and the resultant industrial restructuring.

High R&D expenditure is a recognisable characteristic of thriving economies and there is strong correlation between national expenditures on R&D and competitiveness. Table 4.1 shows total and civil R&D expenditure (GERD)[1] in major countries, as a percentage of GDP.

Table 4.1 Total and civil R&D expenditure (GERD) as per cent of GDP

Country	Total 1991	Civil 1991	Total 1992	Total 1993
Japan	3.05	3.00	3.00	
United States	2.86	2.30	2.81	
Sweden	2.86	2.70		3.11
Germany	2.63		2.50	2.48
France	2.41	2.00	2.40	2.41
United Kingdom	2.13	1.70	2.12	
EC (estimate)	1.96	1.80	1.92	
Netherlands	1.92	1.90	1.86	
Norway	1.84	1.80		1.76
Denmark	1.70	1.70		
Belgium	1.67			
Italy	1.32	1.30	1.31	1.30
Ireland	1.02	1.00	1.08	
Spain	0.87	0.80	0.85	0.85

Source: OECD Main Science and Technology Indicators, 1994

Although the UK's total R&D expenditure shows up quite well in Table 4.1, a high proportion of the UK Government's research budget (45.1 per cent in 1992 and second only to USA) goes to defence, which has questionable benefits to industry (OECD, 1994; Porter, op. cit.). Rothwell and Zegveld (1981) have also drawn attention to the high opportunity cost of defence R&D to economic growth - despite some spin-off benefits from defence-related research.

Table 4.2 shows the international ranking of the top 200 companies by R&D expenditure in 1994. The same group of nations dominates Tables 4.1 and 4.2.

Table 4.2   Top 200 companies by R&D expenditure

Nationality	Companies in top 200	% of top 200
USA	67	33.5
Japan	60	30.0
France	18	9.0
Germany	16	8.0
United Kingdom	12	6.0
Switzerland	7	3.5
Sweden	7	3.5
Netherlands	4	2.0
Italy	4	2.0
Canada	2	1.0
Denmark	1	0.5
Finland	1	0.5
Belgium	1	0.5
Panama	1	0.5

Source: Company Reporting Limited, June 1995

R&D Expenditure in the UK (GERD)

Table 4.3 sets out the constituents of the UK's GERD for 1992 and Chart 4.1 shows who carried out the R&D in the UK for the period 1986 to 1992.[2] The civil non-defence related in 1992 R&D is estimated at 80 per cent of the total.

Table 4.3   GERD and its constituents, UK, 1992

GERD	£12,619m	100% GERD	2.12% GDP
GOVERD	£ 2,032m	16% GERD	0.34% GDP
HERD	£ 2,141m	17% GERD	0.36% GDP
BERD	£ 7,930m	63% GERD	1.33% GDP
ORD	£ 516m	4% GERD	0.09% GDP

Source: OECD Main Science and Technology Indicators 1994/2; HMSO (1993)

After falling from 15 per cent of the total GERD in 1986 to 13 per cent by 1988, Government R&D expenditure increased to 17 per cent in 1991 and 1992. During this period BERD decreased from a high of 68 per cent to only 63 per cent in 1992 whereas the universities' share of R&D (HERD) has risen steadily from 14 per cent to 17 per cent of the national total.

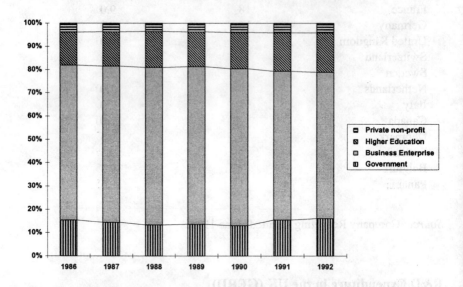

Figure 4.1   R&D by performer as per cent of gross domestic R&D expenditure

Source: OECD Main Science and Technology Indicators, 1994/2

Business Expenditure on R&D (BERD)

Although UK businesses dominate the R&D spending in Britain, 63 per cent of the total in 1992, R&D expenditure by UK business enterprises as a percentage of GDP declined from 1.47 per cent in 1989 to 1.33 per cent in 1992. In real terms, using the GDP deflator, it has decreased by ten per cent since 1990 but as a percentage of sales remained fairly steady at around 2.2 per cent over the period 1985 to 1990, (CSO, 1993a). The annual Company Reporting statistics however show a steady upward trend in R&D investment by manufacturing industry since 1990 and R&D

investment by UK firms is growing faster than in some of the main competitor countries including Germany, the USA and Japan.

Sectoral Differences

Table 4.4   UK Companies expenditure on R&D, 1992

Sector	R&D value £'000s	R&D value as % sales	£'000s per employee
Healthcare	1,550,851	7.68	5.89
Fuel	878,714	0.74	2.58
Chemicals	802,094	4.03	3.69
Electrical & electronics	680,970	3.85	2.23
Food	566,711	0.93	0.84
Aerospace	502,479	3.19	2.54
General manufacturing	493,067	1.06	0.85
Telecommunications	249,957	1.80	1.13
Automotive	238,934	3.01	1.53
Housing & construction	111,950	0.77	0.56
Service industries	106,691	1.90	2.56
Conglomerates	63,768	0.34	0.23
Metals & mining	57,823	0.54	0.50
Water	39,012	0.80	0.71
Electricity	35,600	0.25	0.60
Leisure	34,829	0.26	0.14
Packaging & containers	25,219	0.47	0.43
Consumer products	14,189	0.35	0.13
Publishing & broadcasting	9,399	0.37	0.23
Financial	6,278	3.90	1.42
Retail	3,570	0.80	0.42
Office equipment/ services	3,059	4.13	1.45
All industry composite	6,475,164	1.55	1.55

Source: The 1993 UK R&D Scoreboard, Company Reporting Limited

There are wide differences between industrial sectors. Table 4.4 gives the breakdown, for 1992, of R&D expenditure as a percentage of sales and expressed in £ per employee. Six out of the 22 industrial sectors identified in Table 4.4 accounted for 85 per cent of the estimated industrial expenditure on R&D in 1992 as estimated by Company Reporting Limited. There are also wide variations within industries. In the 1994 figures for example, although all sectors increased their annual R&D investment, three sectors pharmaceuticals, chemicals and healthcare accounted for 96 per cent of the increase and pharmaceutical companies alone were responsible for 86 per cent of the growth.

Company R&D International Comparisons

International comparisons also show significant differences in the mean values of R&D for those companies included in the top 200 classified by R&D expenditure, see Table 4.5, which refers to R&D expenditure in 1992. The 75 US companies in the top 200 spent on average 7.33 per cent of their sales revenue on R&D compared with 5.56 per cent by the 11 UK companies in the top 200. The comparisons of R&D per employee are however more stark with Japanese and American firms spending 2.7 times and 1.76 times respectively as much as the UK firms.

Table 4.5 Mean values of R&D for top 200 companies by R&D expenditure, 1992

	Number in top 200	R&D £'000s employee	R&D as % sales	R&D as % dividends
US companies	75	10.21	7.33	383
Japanese companies	59	15.54	5.86	775
German companies	15	7.07	6.98	1006
French companies	13	6.19	5.56	2102
UK companies	11	5.81	5.86	209

Source: The 1993 UK R&D Scoreboard, Company Reporting Limted

There are some worrying figures for the UK in these statistics. For example, in 1994, only one UK Company in the electronic and electrical equipment sector figured in the top 200 compared with 18 from Japan and 14 from USA. The R&D expenditure by this UK company (General Electric) was £406m compared with £17.8b by the Japanese and £7.5b by the US firms. The four German firms in this sector spent £4.2b and the one Swedish company in the group spent £1.1b. In engineering there was also just one UK company (Rolls Royce) in the top 200 and 13 each from Japan and USA. In chemicals only ICI was ranked with R&D of £184k compared to five German companies which spent £3.9b, eight American firms which spent £3.8b and 13 Japanese firms, £2.6b. Fortunately there are also some encouraging signs, notably in the healthcare and pharmaceuticals sectors in which UK companies have steadily improved their position. In pharmaceuticals, Britain had four companies ranked, spending £2.4b compared with seven Japanese companies spending £1.5b and four American firms which spent £1.8b on R&D. The UK healthcare sector increased its R&D spend by 17 per cent and pharmaceuticals by 11 per cent on the year compared to the international figures of 11 per cent and 5 per cent respectively

Part of the explanation for national differences may be seen in the R&D as a percentage of dividends column of Table 4.5. The top UK firms paid out almost half as much in dividends to shareholders as it spends on R&D whereas the German companies paid out ten times as much on R&D as in dividends. (The French figure for R&D as percentage of dividends is distorted by one company reporting R&D as 17,425 per cent. Excluding this company the mean for the French firms is 570 per cent.) These may reflect differences between the Anglo-American and Continental European views of companies. Writing within the context of corporate governance, but highlighting attitudes which will affect the funds available for R&D and other purposes, Sir Adrian Cadbury (1993) has pointed out that

> The Anglo-American view of a company is that it is based on the capital invested in it by the shareholders. The key relationship in a capitalist enterprise is between owners and managers. Hence our emphasis on the rights and responsibilities of shareholders. The Continental view is much more along the lines of companies being partnerships between capital and labour. They are seen as coalitions of interests and as serving a wider purpose than providing a return to shareholders. Considerable emphasis is, therefore, placed on the relationship with employees in the Continental model and on

the place of the company in the community. This is in contrast to the rights of shareholders, especially minorities, which have traditionally received scant attention.

There is a view that Britain will become increasingly uncompetitive if it remains locked into a financial system that rewards short-term dividend growth and deters longer term investment into intangible areas such as R&D and innovation (Caborn, 1996). It remains to be seen what impact ever closer European Cooperation and the contrasts between the proportions of profits paid in dividends by British and continental companies will have on the attitudes of UK investors and the City of London. A break with current practices and expectations would require a major culture shift.

UK University R&D (HERD)

In 1991/92 the higher education sector carried out 17 per cent of the GERD in the UK, valued at just over £2b. Figures 4.2 to 4.4 show the value of 'externally funded' R&D carried out by UK universities and the sources of that funding which include Research Councils and charities, government departments, industry and commerce, hospital authorities and overseas sources including the European Community. By 1993/94 the total of externally funded university R&D had risen to £1.3b, an increase of 225 per cent in real terms since 1984/85. Adding to this the sum provided for research by the Higher Education Funding Councils brings the total value of research carried out by UK publicly funded education institutions in 1993/94 to £2.1b (CVCP, 1995). In the ten years after 1984, the contribution from the Research Councils and charities, dropped from 58 per cent of the total at the beginning of the period to 54 per cent in 1989 and then increased gradually to 59 per cent in 1994. Work for government reduced steadily from 21 per cent to 13 per cent over ten years while that funded by industry rose from 12 per cent in 1984 to a peak of 14 per cent in 1989 and then fell back to 11 per cent in 1994.

£m

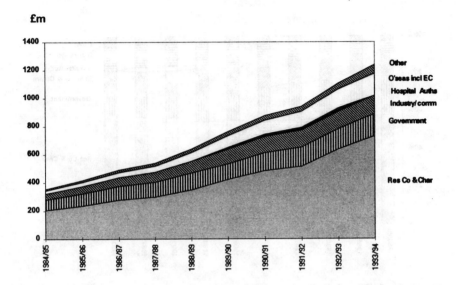

Figure 4.2 Total UK externally funded university research
Source: Universities Statistical Record

The most significant increases in university R&D funding both absolutely and proportionately came from overseas sources, mainly as a result of the universities' participation in successive EU 'Framework' research programmes. Income from these sources more than doubled from 6 per cent to 13 per cent over the decade.

About 20 per cent of the UK R&D expenditure in 1992 was defence-related. Seventeen per cent of the total R&D expenditure (about £2b) was performed by the higher education sector and since relatively little defence-related work is done in the universities, they carry out an even greater proportion of the civil research, about 21 per cent in 1992 (OECD). (For a review of collaborative research in the UK see Handscombe, 1993).

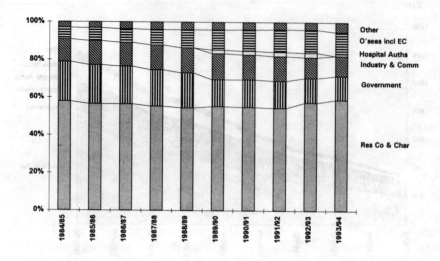

Figure 4.3 Total UK externally funded research by source
Source: Universities Statistical Record

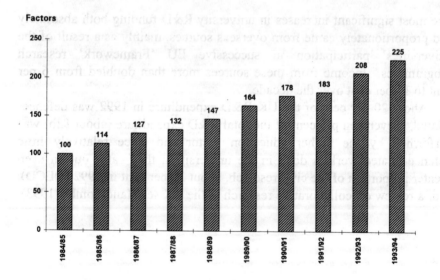

Figure 4.4 Growth in externally funded university R&D at constant prices

Economic Impact of University Research

There is widespread consensus about the importance of university research both to national and local economic development (IACHEI, 1985; HMG, 1993; NIEC, 1993; Porter, 1990; Beveridge and Blair, 1994; CIHE, 1990). The agreed view is that the entire spectrum of university research from pure curiosity-driven research through strategic to near market R&D is of great potential value and that more could be done to exploit it. Nearer market and more applied research is viewed as being of particular value to industrial companies which can collaborate with universities to supplement their in-house research. Despite this the results of a survey, reported in 1990, of manufacturing, process and service companies showed that although 70 per cent of UK companies surveyed had used universities as a limited source of technology, in-house corporate R&D was by far the more important source (CEST, 1990a and 1990b). Universities in the UK have been encouraged by successive governments, most recently through the Foresight initiative (OST, 1995; Kenward, 1995), to become more commercially focused in their research and to collaborate with industry. This message however is at odds with apparently contradictory signals coming from university funding councils which fund university research according to assessments of research quality based largely on peer-reviewed publications. This tension is referred to more fully in Chapter 5. Clearly a compromise is required and most universities are now wrestling with this problem. Individual solutions are many and varied. Walshok (1994), for example, has described a US based university-industry programme in which basic research can be complementary to the expanding concern with technology-transfer and the application of knowledge.

Arguing that universities should go beyond the development of technology-transfer programmes, Walshok (1996) believes that research universities have an increasingly significant role to play if they do three things:

- Embrace a wider and deeper understanding of the multiplicity of factors affecting economic development in a knowledge based society;

- See their role as mobilising and making accessible campus-wide resources from the sciences to the humanities relevant to the knowledge problems confronting advanced economies;
- Invest politically and financially in the development of institutional mechanisms whose central role is to facilitate and develop knowledge across academic disciplines and the boundaries separating the traditional research and teaching programmes from the concerns and challenges confronting the larger society.

Another view is that there is likely to be a mismatch between what university researchers are interested in and what local companies need and that therefore the universities can contribute more to local firms by concentrating on global and international companies as they will then be able to introduce new ideas and concepts (Goddard et al., ibid.). A difficulty with this argument is that the UK is not homogenous and that regional needs and conditions differ throughout the country. What is appropriate for one university in a particular region may be inappropriate in another. Nor need assistance to local companies - much of which is categorised as Other Services Rendered (OSR) as distinct from research and is frequently carried out by industrial units which are decoupled from academic departments - prevent a university's participation in world class research.

The observation has also been made from Canadian experience that universities perform different functions at local and national levels and that as a result links with the local community will be both quantifiably and qualitatively different from those farther afield (Lawton Smith and Atkinson, 1992). In regions where industrial research is low, university research becomes relatively more significant. In Northern Ireland, for example, which has a very small manufacturing base coupled with high unemployment, the two local universities carried out research valued at £41m, almost as much as the estimated £49m private sector expenditure[3] during 1991/92 (NIEC, 1993). Compare this with the figure for the UK as a whole where university R&D of £2 billion was only about a quarter of that carried out by industry.

The impact of the exploitation of university research on a local economy was the subject of a report by consultants Segal Quince and Wicksteed for the Manpower Services Commission (Segal Quince and Wicksteed, 1988). The study, which was based on experiences of the

Universities of Cambridge, Warwick, Salford and Newcastle identified seven broad sets of inter-related factors to be most significant namely:

- The urban and industrial context;
- The culture of the university and the nature of industrial linkage mechanisms;
- The fields and quality of the research being conducted;
- The presence of talented, supportive and motivated individuals and institutions;
- The availability of resources for special initiatives;
- Time;
- Public expenditure.

Collectively the universities have the potential to make a considerable impact both on the national and their regional economies. Both government and the private sector in the UK should be concerned to ensure that the annual £2b investment is well managed, yet its management receives scant attention. The effectiveness of technology-transfer mechanisms, to ensure that the outcomes of research are protected where necessary, exploited commercially and disseminated appropriately, should be of more widespread concern. Too often the disposition of research has been left to the whim of individual researchers. (Although since the ending of the BTG monopoly on exploitation of the outcomes of Research Council funded R&D there have been efforts to ensure its continued exploitation by the recipient universities). Until comparatively recently the value of intellectual property, and incidentally its ownership by the university as employer, was seldom recognised. Career prospects, recognition and credit for academics are still determined largely by their publications, yet to publish may be the least helpful step for the institution, for the economy, or in the long run for the researcher if a discovery has significant commercial potential.

Summary

Although not sufficient by itself, R&D expenditure is a useful indicator of innovation and allows both international comparisons and comparisons between industries to be made. There is a strong correlation between national expenditures on R&D and international competitiveness.

Total expenditure on R&D in the UK is in the region of £12b of which about £8b is spent by Business, £2b by Government and just over £2b by Higher Education. There are wide differences between expenditure by different industrial sectors with companies in healthcare, pharmaceuticals and chemicals spending most on R&D. International comparisons for 1992 showed that the top UK firms come fifth after USA, Japanese, German and French companies in terms of R&D expenditure per employee.

The universities have an important role to play and carry out about 17 per cent of all UK R&D. There has been a steady upward trend in R&D expenditure carried out by the universities, which grew from 14 per cent to 17 per cent of the national total in the period 1986 to 1992. The greatest growth in funding came from European Union (EU) funded R&D, mostly under successive Research Technology and Development (RTD) Framework Programmes (Cullen, 1996).

In peripheral regions the impact and relative importance of university R&D is greatest. In Northern Ireland , for example, R&D carried out by the local universities was almost equal to the total private sector R&D whereas in the UK as a whole university R&D is only about a quarter of that carried out by industry.

Notes

1. GERD is the total intramural expenditure on R&D performed in the national territory, it is the highest aggregate of R&D in the statistical sources and comprises:
    *   Government (Intramural) Expenditure on R&D (GOVERD):
        R&D performed within government R&D establishments including Research Councils;
    *   Higher Education R&D (HERD):
        R&D performed by universities and other higher education institutions;
    *   Business Enterprise R&D (BERD):
        R&D performed by industrial concerns including utilities, services and contract R&D establishments;
    *   Other R&D (ORD): Mainly R&D performed by charities.
2. The Company Reporting figures probably underestimate the total R&D by companies. They are compiled from companies' published accounts up to 31

May 1993 and are based on the R&D funded by companies themselves, excluding externally funded research and take no account of unlisted companies. Also the figures disclosed in companies' annual accounts may differ from those in the CSO figures used by government due to differing definitions of R&D , but they are adequate for present purposes.

3. Business expenditure on R&D in Northern Ireland in 1993 was £39m, equal to 0.43 per cent of the UK total of £9,069m (CSO, 1993b).

5 University/Industry Collaboration

Origins and Development in the UK

For many years university/industry interaction in the United Kingdom was conducted in a somewhat laissez-faire manner. By and large, universities simply reacted to the occasional requests which they received for assistance from industry. Some did so more enthusiastically than others but with few exceptions neither universities nor companies felt motivated to take the initiative in forming lasting relationships. Work for industry was seen by many academics as something of a diversion from their proper role as teachers, researchers and scholars and received low priority. This other-worldliness and seeming disinterest conveyed itself to industrialists, with the result that many in industry, although prepared to recruit graduates, did not approach the universities for help with their practical problems. Few consciously thought of the universities as sources of new technology for their businesses.

By the late 1960s however, unfavourable comparisons were beginning to be made in Britain with practices overseas and particularly with the situation in the USA. There technology-transfer and university/industry collaboration were well developed and university spin-out companies were commonplace, notably from centres such as MIT, Stanford and University of California (see also Bullock, 1983; Roberts, 1991).

Against this background the Confederation of British Industry (CBI) set up, in 1967, a committee of university vice-chancellors and industrialists under the chairmanship of the then head of BP Limited, Mr P Docksey, to study the existing relationships between the universities and industry in the field of research and to make recommendations. The Docksey report, which appeared in 1969, proved to be something of a watershed. It was well researched and identified one of the main difficulties to improved collaboration as being lack of time on both sides. It proposed that those involved in collaboration should be encouraged to spend 10 to 20 per cent of their time on this activity. The Docksey Report preceded innovations such as science parks and teaching company schemes and serves as a useful touchstone (CBI, 1970).

Following Docksey there was a flurry of activity. Universities began to employ full-time staff to deal with their relationships with industry. Writing in 1977, Smith reported 33 universities which regarded industrial liaison as an integral part of their activities (Smith, 1977). The next major impetus was unquestionably the government funding cuts of 1981. Some institutions suffered more than others but virtually all began seriously to review their procedures and mechanisms for dealing with industrial collaboration (CVCP, 1981). John Ashworth, then Vice-Chancellor of Salford, a university which suffered more than most in the cutbacks, drew attention - in his Redfearn Memorial Lecture in 1984 - to the need for universities to develop appropriate organisational structures to facilitate good linkages with industry (Ashworth, 1984). Salford University was one of the first to create structures aimed specifically at making it attractive as a partner to industry and Salford went on to become one of the most commercially oriented of UK universities.

A report by the Advisory Council for Applied Research and Development (ACARD) in 1983 attracted a great deal of interest not least from the, then, Prime Minister Margaret Thatcher. It is a well balanced document which lists the various roadblocks to successful cooperation between industry and higher education institutions (HEIs) and makes a number of very sensible proposals including the important point that industry would need to match the initiative taken by the HEIs with an initiative of its own (ACARD, 1983).

By the mid-1980s the HEIs had realised that the need to obtain industrial funding was no polite invitation from government. It affected their very survival. Determined efforts were made by the HEIs to respond to the criticisms which had been made and address the issues raised in the various reports. Indeed the universities felt themselves to be under considerable pressure to put their houses in order (CUA, 1984; Burnett, 1985; ACARD, 1986).

Additional professional staff were recruited, many from industry, to direct university industrial liaison offices and to bring fresh ideas and new perspectives to their roles. The traditional practice of responding to requests for help from industry was replaced by a pro-active approach in which universities set out to market themselves as a resource to industry.

The University Directors of Industrial Liaison (UDIL), which was formed in 1969 as a network of industrial liaison professionals to assist university/industry collaboration, played an important part by codifying and disseminating best practice (Webster, 1988). University industrial liaison staff learned how to deal much more expertly with contractual matters, including the management and exploitation of intellectual property, confidentiality concerns and the costing and pricing of work for industry as well as with project management.

That process has continued and now most universities in the UK employ professionals to deal with their collaborative activities with industry. Almost all have established industrial liaison offices, centres, units or bureaux, frequently known by acronyms to make them more easily identified and easily remembered, for example AURIS at Aberdeen University, ULIS at University of Leeds and UNILINK at Heriot Watt University. Some universities have formed companies to which they delegate the industrial liaison role, more will be said about these later.

The UDIL network grew steadily, promoting the universities nationally as a service to industry. The 1993 UDIL directory of services provided contact names and details of the services offered by 67 universities throughout the UK and the Republic of Ireland (UDIL, 1993). Private publishers also began to produce comprehensive directories of university services, for example, 'Higher Education Resources for Industry', published and updated annually (Hobsons, 1989/90).

Concurrently the polytechnics in Britain were also developing links with industry and the Academic Industry Links Organisation (AILO) was formed in 1979. Although as a group the polytechnics carried out much less research than the universities, many developed close linkages with smaller companies, helping them to improve their technological strengths, through for example consultancy and problem solving. There is a view that some small firms found the polytechnics to be more approachable as sources of assistance and less intimidating than the universities (AILO, 1990 and 1992). With the ending of the so-called binary divide in 1992 the former polytechnics formally became universities and UDIL and AILO, have merged to become the Association of University Research and Industrial Liaison (AURIL).

One result of this activity has been a steady increase in industrially funded research which increased from £42m to £132m in the period 1984 to 1993, i.e. more than doubled in real terms (Universities Statistical

Record, various years). Government funded programmes such as the Alvey programme for information technology and subsequently a range of schemes such as LINK and more recently - following the white paper 'Realising our Potential' - ROPA Awards, have further stimulated togetherness between academia and industry. Successive European science and technology research Framework Programmes have also favoured consortia involving companies, universities and research establishments throughout the European Community.

It should be noted however that although UK universities have mostly responded enthusiastically to EC programmes the net benefit of this participation has been questioned. There are those in industry who doubt whether the benefits of such research feeds through to UK industry. A study by Pike and Charles (1995) suggests that there are both positive and negative outcomes from participation in EU research programmes. The latter spring from the partial funding of EU research which often barely covers the universities' marginal costs, the money and time spent in travel throughout Europe to attend reporting and monitoring meetings and frequent difficulties over intellectual property arrangements.

Other types of collaboration, including the provision of specialised services and access to university facilities, have also increased steadily. The media have reported numerous examples of successful transfer of technology from the universities to industry by means of contract research, consultancy, problem solving, teaching company schemes, science park developments and staff transfers. Significantly, many of these stories have been provided by the HEIs themselves to publicise their successes - evidence that universities felt the need to demonstrate publicly their industrial relevance (Financial Times, 1984a, 1984b, 1984d; Glynn, 1984; Cookson, 1989a; Ditzel, 1990; Handscombe, 1991).

Although most universities can, and increasingly do, point to successful collaboration and many industrialists testify to mutually beneficial joint activity, it is disappointing that there is so little in the literature during this period of rapid change giving industry's point of view. Writing in the Financial Times in 1984 Fishlock referred to university/industry interaction as being too often a dialogue of the deaf (Financial Times, 1984c).

In 1986 an independent body, the Council for Industry and Higher Education (CIHE), was formed under the chairmanship of the Rt. Hon. James Prior MP, then Chairman of the General Electric Company. This group comprised twenty-two heads of large companies and ten vice-chancellors, polytechnic directors and heads of colleges. The Council's stated aim was to encourage industry and higher education to work together and to represent their joint thinking to government. One of the group's first actions was to address some of the difficulties inherent in university/industry collaboration. Its first report emphasised the importance of increasing research technology-transfer and offered the opinion that companies will be unsurprised to see their own and government's research money spread unequally among individual universities and departments (CIHE, 1987a and 1987b). Had it been fully accepted by government and incorporated into policy, this view - with its implication that some universities might be starved of research funds and forced to concentrate only on teaching - would have considerable implications for HEIs. Indeed subsequent events such as the new mechanisms for awarding research funds based on assessments of research quality are beginning to have this effect.

The Foresight exercise which began as a result of the UK Government's white paper 'Realising our Potential' has stimulated university/industry collaboration in a formal way through joint participation in sectoral panels, for example Campbell (1994). So too have successive EU Framework Research and Technology Development programmes, most of which require research consortia comprising industry and academic researchers from a number of EU member states. Although generally both industry and academia recognise that there are mutual benefits to be had from working together, there are dissenting voices in both camps. Some in industry recognise that there are fundamental differences in the value systems in business and academia and accept that academic and industrial values should be different. In these circumstances attempts to force the two sides together is to the longer term benefit of neither (Robinson, 1995).

It has also been argued that local universities contribute most effectively to their local economies not through their research efforts - the results of which should be made freely available to all companies - but by delivering the right innovating graduates which their local companies need (Beckers, 1992). Yet another point of view is that the real returns to the universities from collaborative research are primarily intellectual, not financial and that academics can broaden their intellectual horizons by relating their work to others' problems CISAT (1991). These differences in perception underlie many of the current issues and frustrations being experienced both by universities and industrialists.

Some Current Issues

Costing and Pricing of External Work

One of the most contentious issues bedevilling university/industry collaboration in the UK surrounds the costing and pricing of externally funded work. Although quite separate activities, costing and pricing are frequently confused and muddled - usually through ignorance or naïveté by those who do not appreciate the difference but sometimes deliberately by those who do.

For many years universities had been exhorted by governments to charge, not only for the direct costs of external work but for a contribution towards their overheads. This contribution, for long set at a minimum of 40 per cent of the direct costs, was to help pay for those indirect costs which were not easily separated as being identifiable with a specific piece of work.   It was intended to contribute for example to general administration costs, heat and light, professional indemnity insurance and the depreciation costs of equipment and premises. In practice the 40 per cent overhead rule, intended as a minimum, became a starting point for negotiation and most contracts were settled at much lower levels. Although UDIL made efforts to persuade the universities to close ranks and hold out for higher fees for external work and services provided, many industrial liaison staff were unable to prevent their academic colleagues from offering their services at lower rates in what they saw as competitive situations.   Not surprisingly, companies and also many government departments exploited the situation to force prices down. In an effort to return to first principles and devise a costing method which would be familiar to and therefore more acceptable to industry, the Committee of Vice-Chancellors and Principals (CVCP) commissioned a study under the chairmanship of Professor H J Hanham, then Vice-Chancellor of Lancaster. The Hanham Report, which appeared in 1988 after widespread consultation with industry and government, proposed a methodology for the costing and pricing of external contracts by which overheads would be calculated and levied as a percentage of the staff costs incurred on projects

(CVCP, 1988). The Hanham principles were adopted widely and most UK universities have since made determined efforts to obtain better prices. However, despite the prior consultation and welcome in principle which the Hanham Report received, it has proved to be very difficult to overcome years of under-charging and the widespread perception that because universities are publicly funded, their indirect costs are somehow less real than those of their clients.

The problem has been recognised by industry in the company response to the 'Towards a Partnership' paper referred to above. The industry view is that there is scope for much more joint research and that this would be encouraged by resolving different views on costing, pricing and exploitation of intellectual property rights (CIHE, 1988). A common difficulty is that the clients are often in a stronger bargaining position than universities whose researchers are often desperate for funds and although the situation is slowly improving, UK universities are still subsidising industry and government departments on most collaborative projects. Paradoxically, the greatest resistance to obtaining realistic contractual terms often comes, not from potential clients, but from university researchers who undervalue their own worth. Many academics are fearful of losing what are still referred to as 'grants' from industry, a term which implies some kind of charitable or altruistic donation or sponsorship rather than consideration for a commercial contract through which both parties expect to benefit. As a result most university researchers fail to ask for commercial terms for work which they undertake for industry and government. This may be a legacy of attitudes more appropriate to a time when universities were funded more adequately both by government and by wealthy sponsors and patrons.

Part of the trouble is that a great many academics do not understand the economic facts of life. Most think in terms of recovering only the marginal costs of work they do for others and few understand the concept of opportunity cost. It is of course easy to argue that, if there are few alternative opportunities to forego, this does not matter - academics whose expertise is in demand can hold out for better prices in contrast to those seeking funds, who are inclined to accept whatever is offered. By doing so they perpetuate the view in industry, and even more particularly in government, that the universities are a cheap source of expertise. The dilemma is how to break out of this cycle of low expectations on both sides.

Despite the intervention of industrial liaison professionals to stiffen the resolve of academics, many researchers are apologetic about asking for what they privately feel to be unrealistic charges. Almost invariably therefore they are offered lower prices than those which they have been advised to quote. Most are either too faint-hearted to argue or lack the ability to present their costs in the context of potential benefits to the client and they interpret the situation as confirming their personal belief that the asking price was too high. The result is that their expectations do not increase and the process repeats itself.

The challenge facing universities is to break out of this cycle by encouraging their researchers to have increased expectations and re-evaluate their worth to industry. To achieve this it may be necessary to refuse contracts which do not meet minimum commercial criteria and impose restrictions on the freedom of staff to negotiate contractual conditions unaided. This could be difficult to enforce and would no doubt be viewed by some members of staff as striking at the very roots of their academic freedom to do what they like with what they regard as their own research. It is worth noting that this difference in perspective between university industrial liaison staff and academics is not confined to the UK. Rahm (1994) reports on a similar dichotomy in USA. If they are to become more 'commercial' in contractual matters and if they expect their clients to pay economic fees for their services, universities must demonstrate their ability to manage collaborative projects effectively and to complete them on time. This is a lesson which academia has been slow to learn and one to which the CBI and CIHE drew attention in their response to the draft Hanham Report (CVCP, 1988).

Ownership and Exploitation Intellectual Property

There is a growing feeling in the universities that they are in a 'no-win' situation. Universities have long been criticised both by government and by industrialists for being ivory-towers, unaware of commercial realities. Yet now that they have become much more commercial and assertive of their rights, they are facing a reaction from firms reluctant to pay economic costs for their services and access to their intellectual property.

The issue of the ownership and subsequent exploitation of intellectual property arising from industrial funded research is closely linked to the pricing debate and properly viewed by the universities as an inclusive part of the contractual terms. Universities, almost universally, now refuse to assign away their background intellectual property which has often been acquired over many years and has usually been funded from a variety of sources. Foreground IPR however i.e. that arising from a current piece of work, is treated differently. Put at its simplest most universities take the view that if full economic costs are paid for a piece of research then the sponsors have rights either to free or exclusive use of intellectual property which results. When, as is frequently the case, less than full costs are paid, universities generally wish to retain ownership of the IPR or at least expect to negotiate realistic royalties from the commercial exploitation of work which the universities have in effect subsidised. The university position was clearly laid out by UDIL and subsequently by AURIL in guidelines for universities (UDIL, 1988; AURIL, 1995).

Many companies, on the other hand, argue that this is a simplistic view and that universities underestimate the additional work and costs involved in developing a research idea into a marketable product. They claim that it is industry which is best placed to exploit IPR and that they require a free hand in order to do so.

Taken to extremes these are irreconcilable positions and in reality compromises are usually reached but the issue continues to generate a great deal of heat and strong feelings. From time to time the debate erupts, for example in Research Fortnight during the first half of 1995 when both sides aired their views, see for example Elliott (1995) and Harvey (1995).

Government departments too, which as a sector are heavy users of university research and services, often prove difficult to convince although the situation has been helped by the more liberal regulations about IPR incorporated into EU contracts. These allow those members of research consortia, such as universities, which have not got the ability to exploit IP themselves to receive an appropriate royalty from those which have.

The value to industry of being able to license-in new technology is widely acknowledged, see for example Drew and Edge (1985), yet the protection of intellectual property, frequently through patents, and its subsequent licensing-out varies widely throughout the UK. A survey reported by Packer (1994) revealed that the total number of patents held by respondents from the old (i.e. pre-1992) UK universities was 510 and the maximum held by any university was 60. Although there are many examples of know-how and non-patented IPR being licensed to industry - and indeed in fast moving technologies such as consumer electronics patenting is much too ponderous a procedure to contemplate - these figures are disappointingly low. By contrast MIT makes about 380 disclosures resulting in 105 to 120 US patents each year and grants 80 new licenses annually (Nelsen, 1994).

It is perhaps unfair to make comparisons with MIT which has a unique record of transferring technology and whose reputation is such that it is actively sought out by industrialists as a source of innovation and ideas. This reputation was not acquired by accident however and there is much that could be learned from MIT's experiences especially since intellectual property issues are generally regarded as being of greater importance in the USA than in Europe (Williamson, 1992).   Some of the difficulties experienced in the UK may be similar to the differences in perspectives of academic administrators and researchers noted in the USA and reported by Rahm (1994). In a follow up to her earlier survey Packer (1995) compared and contrasted the views of industrial liaison staff and researchers in Britain and concluded that there is a need to reappraise the role of patenting in UK universities.   Importantly, Packer states that this reappraisal is needed not with a view to identifying 'best practice' which she argues fails to take account of differences between institutions.

There is an obvious need to improve mutual understanding between industrial liaison professionals and researchers about issues surrounding patenting as well as commercial exploitation of IPR.   One of the recommendations in a recent report by the National Academies Advisory Board was for a better understanding of the commercialisation process and the role of intellectual property rights amongst the academic community. Additionally universities were urged to consider actively how far a knowledge of intellectual property should become a part of undergraduate and graduate courses (NAPAG, 1995, para., 11.4). However, given that only a small percentage, variously estimated at about 5 per cent, of patents result in income for their inventors it is not surprising that many university administrations are luke-warm about patenting. The costs of acquiring a patent portfolio and especially of having to defend a patent, are not inconsiderable and in their present straitened circumstances it is probably right that universities should insist that these costs should have to be justified. Of course those universities which have been successful have less difficulty in doing so but for many the analogy with the National Lottery is hard to dismiss.

Pressures from Government

Another frustration has been that although universities are urged to be commercial and more customer-oriented - more like industry - they are also being urged, by government as well as by some in industry, to help smaller firms at less than economic costs. Universities have been criticised for not marketing their services more vigorously to smaller companies (Fielden & JM Consultancy, 1991). This however ignores the reality that, as in so many business activities, a variation of the Pareto principle applies. Most R&D is carried out by the relatively few larger companies and the much more numerous smaller companies are frequently unable or unwilling to pay realistic prices. If the universities really were to behave like commercial companies they would ignore many of these potentially less profitable activities and concentrate their scarce resources on the 20 per cent or so of key customers which provide the bulk of their external earnings. The fact that they have not taken such a purely commercial view is evidence that most universities accept that they have some social role or responsibility to assist with economic development. This sense of responsibility, which in the long-term can be viewed as enlightened self-interest, is apparent in Northern Ireland where the economy is weak. There both universities have publicly expressed their commitment to collaborate with government development agencies to help stimulate industrial investment and regeneration through technology-transfer (Beveridge, 1991).

A relatively new tension has arisen since the introduction of new mechanisms for funding research in UK universities by which the various university funding councils allocate funds based on assessments of research quality. Successive research assessment exercises (RAEs) have largely been based on publications in peer-reviewed journals with little more than a nod being given to industrially relevant research. There is no doubt that the research assessments have concentrated minds in the universities but they have also caused some less beneficial effects. In the present context they have discouraged some academics from undertaking commercially funded research which does not 'rate' in the RAEs. Even before the introduction of the new funding procedures it was recognised that peer-reviewed funding, which in the UK is often synonymous with Research Council funding, did not sit easily alongside more applied research. In a forthright and amusing manner Skinner (1989) gives an inventor's perspective of the commercialisation of university-based research. He makes the point inter-alia that a researcher being funded by commercial firms is unlikely to be able to attract contemporaneously research council funding unless he diversifies his research, which itself carries dangers.

Another phenomenon has been the inevitable emergence of league tables, based on the results of the RAEs, of research excellence in various subjects. This may well have beneficial effects in terms of motivating universities to improve their relative rankings but it has also introduced a kind of transfer market through which sought-after academics are being lured from centres of excellence to bolster the ratings of less well endowed universities. There are implications in this for the universities' intellectual property. As Skinner (op. cit.) points out, although one of the major factors affecting senior university appointments is the ability of the candidate to attract research funding, he is restrained from offering this critical component of his worthiness because the intellectual property on which his expertise is based is owned mainly by his present employer.

Developments Elsewhere in Europe

Practically all universities in the UK and increasingly also those elsewhere in the EU, now have professional staff to deal with university/industry liaison. The European Industrial Research Management Association (EIRMA) produced a series of recommendations to encourage and stimulate university/industry cooperation, (EIRMA, 1988). A study, commissioned by the EC under the SPRINT programme, examined the role and activities of industrial liaison officers (ILOs) at universities in all 12 member states of the Community.

Kuhlmann characterised the development of industrial liaison activities in the various EC member states as 'weak' in Greece, Italy; 'emerging' in Spain, Portugal, Ireland, France and 'mature' in Belgium, Netherlands, Denmark, Germany, UK (Kuhlmann et al., 1990).

German universities and industry have for many years had close collaboration through the Fraunhofer-Gesellschaft Institutes and Fachhochschulen (Cheese, 1991). The latter have been particularly successful in building a rapport with SMEs by engaging in applied research (Gering and Schmeid, 1992; Berggreen, 1994).

University/industry collaboration is now well developed in Denmark, and has some novel structures such as a scholarship scheme for innovative entrepreneurs and a National Institute for Product Development based at the Technical University of Denmark. The Institute is a self-supporting commercial venture run by graduates and postgraduates (Danish Technological Institute, 1994; Sheen, 1993). The Swedish universities are particularly active in encouraging entrepreneurship and spin-out companies (University of Linköping, 1992) and the University of Twente in the Netherlands has also pioneered entrepreneurial activity by university students (Kobus, 1992). The developing situation in Portugal was summarised in papers by Fernandes and Sellar (1990) and Pereira (1992). The position in the Republic of Ireland is now almost indistinguishable from that in the UK. For an overview of technology-transfer activity there see McBrierty and O'Neill (1991a); O'Neill and McBrierty (1992); Dineen (1995).

The services being offered and the means through which technology-transfer occurs in the EU countries parallel those employed in the UK and indeed the universities in the OECD highly-developed regions all use variations of the same basic strategies, adapted to suit their own national conditions (Barden, 1993). Attitudes to costing and charging for services vary and are influenced by differing university funding regimes, by legislation and by attitudes of sponsors to ownership and exploitation of intellectual property (Gering and Schmied, 1992; Smith and Atkinson, 1994).

The importance of enhanced collaboration between industry/university partnerships for European growth and competitiveness was highlighted amongst the key conclusions of a report by the Industrial Research and Development Advisory Committee of the European Commission (IRDAC, 1994). Indeed international dialogue and interchange of best practice between those engaged in university/industry activities throughout Europe is steadily increasing so that the prognosis is that, eventually, residual national differences will be solely due to local legislation and funding mechanisms (TII, 1990).

As Europe expands it is predictable that university/industry collaboration will become more widespread and because the developing countries can in effect 'short-circuit' the learning curve they can be expected to catch up rapidly with those countries which only a few years ago were leading the way. Today the situation in several of the countries described as 'emerging' in 1990 is almost indistinguishable from those categorised as 'mature'. University/industry collaboration is already spreading to Eastern Europe and to the Far East, for example to Russia (Morrison and Struthers, 1995), Poland (Dietrich and Kurzydlowski, 1993) and Malaysia (Ratnalingham and Singh, 1993). For a wider international perspective see McBrierty and O'Neill (1991b).

Services Offered and Means of Delivery

Most UK universities and many elsewhere in Europe provide a wide range of services and many have a diversity of delivery mechanisms. The technology-transfer services vary between universities but UK institutions now offer most of the following:

- Consultancy and problem solving;
- Collaborative and contract research and development;
- Participation in EU R&D programmes;
- Provision of specialised training courses;
- Licensing out of intellectual property and know-how;
- Staff secondments / personnel exchanges;
- Provision of and participation in Science/Technology parks;
- Student-based projects at undergraduate and postgraduate levels;
- Teaching Company schemes;
- Academic entrepreneurs and university companies.

It is now recognised that there is no unique best way for universities to collaborate with industry and most universities have developed a range of delivery mechanisms in response to demand from industry (Blair, 1990a; CIHE, 1990). At one extreme, student projects are a low cost way of working with even the smallest companies. Many undergraduate degree courses now require students to undertake practical projects in their final year and universities welcome approaches from companies with projects or problems which lend themselves to this approach. Longer term activities can be carried out by postgraduate students. In the UK, Government support and partial funding is available under the Teaching Company and CASE award schemes. These programmes are therefore particularly cost-effective ways to combine learning with longer term industrial projects.

There is no obligation on the part of the host company to employ students funded under these schemes but almost invariably the opportunity to recruit someone well versed in the company's particular technology and familiar with its problems is too good an opportunity to miss and most students are offered employment.

In most universities academic staff offer consultancy and problem solving in their specialised fields. Consultancy is perhaps the traditional method of importing expertise to a company. It is a relatively low cost option as, usually, the consultant is required for only a short time. Where longer term expertise is required it is generally more economic either to provide training for existing company staff (perhaps by the consultant) or to recruit staff with the necessary expertise. Some academics are heavily involved in consultancy and those who become recognised as experts in their fields can command quite high fees.

Longer term collaborative and contract research is usually best performed by academic departments. For larger projects or those which involve consortia, as is increasingly the case with EU projects and some government funded projects, there is often a need to appoint a full-time project manager.

It is sometimes said that universities are unable to respond speedily to requests for help from industry, a perception resulting partly from unrealistic expectations on the part of industrialists. Universities are sensitive to this criticism and most have now adapted to cope in various ways. It is easy to understand that members of academic staff with teaching and other commitments are not always able to respond instantly to requests for help. Nor is it always possible for them to undertake long periods of industrial work. It is in this type of activity that full-time industrial units and centres are of greatest benefit and many universities have such groups. These units are frequently staffed with people who have no teaching commitments and are therefore decoupled from the normal academic timescale.

Many universities have become recognised as centres of excellence in particular fields. This often occurs when, over a number of years, understanding of a particular topic develops to the point where new knowledge is continually being created. Typically, once a critical mass is reached, this expertise begins to attract interest and funding from a variety of sources, including industry, leading to research of world class.

Some universities have formed industrial clubs to undertake research in particular areas. One such is the Questor Centre at Queen's University Belfast which has a membership of 15 international companies. The Questor Centre was based on an American model - the University/Industry Research Consortium (UIRC), the first of which was established in 1978 to carry out research into applications of composite materials to industrial products (Steiner and Kucich, 1995). Member firms pay an annual subscription in return for which they collectively set the research agenda for interdisciplinary research on environmentally related topics.

Practically all universities now provide a point of first contact for those wishing to use their services. Typically these provide sign-posting and referral services for industry to the most appropriate sources of assistance, deal with licensing and contractual matters related to contracts. Frequently these industrial liaison offices are staffed, not by academics but by people with industrial experience.

Some universities have been notably more successful in their interactions than others. In 1995 the Engineering and Physical Sciences Research Council (EPSRC), commissioned a report aimed at identifying the factors, policies and practices used by those universities which have achieved greatest success in working with industry, in commercial exploitation and technology-transfer (Martin, 1995). Tables 5.1 and 5.2 below refer to work by Martin (1995).

Table 5.1   Success factors from the universities' perspective

Success Factors from the Universities' Perspective
High quality research and training
Multidisciplinary approach to organisation and problem solving
The challenge of real industrial problems
A duty to support society and the economy
Employment opportunities for students
The need for additional resources
Delegated financial authority
An effective Industrial Liaison Office
Academic entrepreneurs
A large concentration of science, engineering and technology
Academic freedom
Networks of contacts, locally, nationally and internationally
Product champions in collaborating companies
The industrial club concept
Staff interchange
Multifunding to provide research and training capability
Consultancy and technology-transfer schemes as 'tin-openers' for future research and training partnerships
Conferences, special meetings
The ability to respond quickly to requests from industry
Generous staffing levels
Charter, mission
Ethos, culture, teamwork, timescales
Managed programmes - monitoring and evaluation
Research Council and DTI industry oriented schemes
State-of-the-art equipment
European Commission Schemes
Career paths for skilled research assistants
Presentational skills, including videos
National agencies: Welsh Dev Agency, Scottish Enterprise, IRTU
Large concentrations of research students
Industry oriented centres e.g. Rolls-Royce UTCs
Minimal bureaucracy

Source: Martin (1995)

Table 5.2   Success factors from industry's perspective

Success Factors from Industry's Perspective
High quality research and training
An emphasis on multidisciplinary culture and organisation
Financial and intellectual gearing
Teamwork and personal characteristics such as loyalty, ability to get things done and attention to timescales
Sound planning
Staff interchange
European Commission Schemes
Government(Research Council, DTI, DoE) sponsored schemes
Networks of contacts - conferences
The industrial club concept
Concentration of resources - teams of research students
Monitoring and evaluation
Minimal bureaucracy
Aid to recruitment
A need to retain in-house research capability as an aid to sound decision making

Source: As Table 5.1

Using as a success rating the ratio of Non-Research Council to Research Council earnings, Martin identified 12 UK universities which had been most successful in exploiting their technology commercially. As well as visiting and interviewing staff in the selected universities Martin obtained information from a number of companies, the DTI, the CBI and the Royal Academy of Engineering. The success factors, arising from this study, which are seen to aid and motivate university/industry collaboration from both an industry and a university perspective are given in Tables 5.1 and 5.2.

Present Significance

It has been argued that the universities are undergoing a second academic revolution, that the character of institutions is changing and they are adopting a more pivotal role in society, both in conjunction with industry and through the exploitation of their knowledge-base (Constable and Webster, 1990; Webster and Etzkowitz, 1991). How enduring these changes will prove to be remains to be seen but certainly in the UK there have been marked shifts in the attitudes both of university administrations and academics to the imperative to collaborate more closely with industry. The spur to this, certainly since the early 1980s, has been successive funding cuts and threats of cuts. One of the effects of which has been to encourage the universities to find innovative ways of earning money from non-government sources. Attempts to force what sometimes seem like shotgun marriages between academia and industry - most recently through the Foresight Programme - have been viewed by many on both sides with some cynicism.

The exploitation of university research through the formation of new companies, which is discussed more fully in the following chapter, is perhaps the most recent addition to the repertoire of ways of transferring technology. Given the frustrations experienced in 'traditional' mechanisms of collaboration e.g. difficulties in obtaining economic prices from industry, the marginal funding arrangements prescribed for many government and EU-funded projects, the contradictions in the messages being received from government, and their own lack-lustre performance in licensing-out their technology, it is not surprising that many universities have decided to be more like industry and have begun either to form companies themselves or to encourage their researchers to do so. By effectively 'decoupling' the commercialising of their research in this way many of the frustrations disappear but of course new challenges arise. These are addressed in later Chapters.

Summary

Beginning in the 1970s and stimulated by government funding cuts in the 1980s something of a renaissance occurred in UK university/industry relations which had until then been carried out by the universities in a purely reactive and somewhat laissez-faire manner. By the mid-1980s practically all UK universities had begun to view their relationships with industry in a new light and as a source of additional funding for research and development as well as simply potential employers for their graduates.

Most had put in place industrial liaison offices and recruited professional staff, frequently from industry, to stimulate and manage collaboration with business and industry. Since then work for and in collaboration with industry has burgeoned and is now accepted as a normal part of university activity. Both UK Government and European Union R&D programmes have further stimulated linkages between academia and industry so that industrially funded research in UK universities increased steadily from £42b in 1984 to £132b in 1993.

Currently the most contentious issues centre on contractual issues related to the costing and pricing of external work and the ownership and exploitation of IPR. Universities are now more determined that hitherto to obtain realistic prices for their services and for access to their intellectual property. They are no longer prepared, for example, to assign away patents and know-how based on background expertise acquired from many years of research at less than economic costs or without realistic expectations of royalty income.

Conflicting messages from government about the need to be commercially relevant whilst government funding for research is based on peer-reviewed research assessments is causing tensions and some anxiety both in the universities and in industry.

Developments throughout Europe have paralleled those in the UK which has in most respects been in the van of developments. With increasing exchange of best practice and European collaboration, stimulated by successive EU framework programmes, the differences in practices between the EU countries are narrowing. Some countries, in particularly the Netherlands and Sweden have been particularly successful in the formation of spin-out companies.

The range of services offered has also grown rapidly and all universities now collaborate in various ways. These include access to and use by industry of university resources and equipment; student projects; consultancy and problem solving; contract and collaborative R&D and equity sharing joint ventures with industry through new company creation. There is now also a great diversity of means by which services are delivered to industry. As well as the 'traditional' academic departments universities now offer services through a wide range of industrial units, some employing their own full-time staff on industrial work; various forms of industrially funded research clubs and centres and schemes such as the DTI teaching company and CASE awards are also widely used. Recently attempts have been made to identify the factors which characterise those universities which have been most successful in obtaining industrial funding. From both the viewpoint of industry and the universities these include high quality research and training and an emphasis on a multidisciplinary approach.

The difficulties and frustrations experienced by universities in the traditional means of collaborating with industry have led many to seek new and innovative ways to exploit their technology. Amongst these is the formation of NTBFs either by the universities themselves or by university staff.

6 University Companies

Background and Developments in the UK

A comparatively recent feature amongst universities in the UK has been the emergence and rapid increase in the number and variety of enterprises formed to exploit university research and development. These vary enormously in type, objectives, method of organisation, and extent of university involvement.

In a report prepared for the Committee of Vice-Chancellors and Principals (CVCP), in 1986, Lowe recorded that about 20 universities in the UK had formed holding companies or structures of various degrees of sophistication as vehicles for company formation. Most of these were engaged in contract research or the provision of consultancy (Lowe, 1988). Lowe also identified a much larger group of more than 500 'spin-out' companies emanating from universities in the United Kingdom. These he defined as companies which operated independently from the university or in which the institution had either no holding or only a very minor stake and little hands-on involvement. These companies varied in size and degree of maturity from embryonic firms employing only a few people to well-established enterprises with sales turnover of £ millions. This latter group, which appeared to be much more dynamic and successful than the university-owned enterprises, was also mainly concerned with consultancy and research but about a third were engaged in product assembly or manufacture.

Since Lowe's survey many more companies have been formed both by universities themselves and by academic entrepreneurs acting with various degrees of support from their institutions. The stimulus to universities to exploit their research commercially has come from government. The Higher Education Funding Councils (HEFCs), which came into being in 1992/93, have made much clearer the distinction between their funding for teaching and that for research. Those institutions whose research is most highly rated not only receive more direct government funding but find it easier to attract additional external funding from sponsors such as Research Councils, EU research programmes and other sources. For those universities which score less well an ever greater proportion of their income will have to come from non-government sources and the encouragement of entrepreneurship has been viewed by many as a way to achieve that end. Further impetus to entrepreneurship has been provided by government initiatives to encourage an enterprise culture and stimulate entrepreneurship. An aim in development regions has been to combat unemployment by encouraging self-employment and ultimately to grow more indigenous firms, see for example IDB (1985). One such scheme was the 'Innovation Initiative' introduced by the Department of Trade and Industry (DTI) to encourage the exploitation of research (DTI, 1990).

The European Union (EU) has helped to fund Business Innovation Centres (BICs) which bring together, in partnership, local authorities, universities and the private sector to encourage and stimulate enterprise and new company formation at the local level. A number of such BICs have been formed in the UK, see for example King (1990). The net effect of these programmes has been to tempt universities as well as erstwhile academic researchers on the road to entrepreneurship. A number of universities have chosen new company formation as their favoured route to commercial exploitation of their research, others have added this method to their repertoire of ways of exploiting their technology and make conscious choices, depending on the circumstances, between company formation and other more traditional methods of exploitation.

There has been a good deal of informal contact and exchange of experience amongst universities which have formed companies. It was a fairly natural development therefore that, in April 1994, a University Companies Association (UNICO), was formed with the stated objectives: 'To arrange meetings of Members, to facilitate interaction between Members and with third parties and to promote their common interests'. Thirty-three university companies were present at the inaugural meeting. They were however a disparate group ranging from companies which exist solely to sell their universities services to those which are holding companies for a number of high-technology product-based firms.

Developments Elsewhere

Europe

Beyond the UK, the growth of university spin-outs in Europe has been greatest in Scandinavia - notably in Sweden and also in the Netherlands, with recent increases in Germany and Belgium. There is intense interest however in all of the EU countries and technology-transfer is increasingly being viewed as a means of accelerating economic growth.

A review by Oloffson and others of technology-based new ventures from Swedish universities revealed that new companies were emanating there at an increasing rate from seven institutions. During the period 1980-82, 27 significant companies - defined as those having a turnover of 100,000 SEK (about £10,000) or more in their second year of operation - were founded compared with 63 during the period 1983-85. There were many more smaller firms. Oloffson estimated that the rate of company starts by university researchers in Sweden during the 1980s is at least on a par with that of Cambridge (Oloffson et al., 1987).

A survey of companies with origins in the high-tech environment of Linköping, Sweden, published in 1992 by the Industrial Liaison Office of the University there describes the activities and essential operating statistics of 53 spin-outs from that university and from other technology-based companies in the region. More than half of those described emanated from the University of Linköping. Thirty-four are owned by their chief executives, their founders or employees and 12 are wholly owned by outside investors. The companies range from single-person consultancies to the largest which designs, develops and markets computerised information systems, employs 100 staff and has a turnover of MSEK 150 (about £15m). Median and average turnover of these spin-outs was MSEK 5 (£500k) and MSEK 12 (£1.2m) respectively. Taken together they provided 650 full-time jobs and contributed MSEK 630 (£63m) to the local economy (University of Linköping, 1992). An important point is that, although it provides a supportive and encouraging environment, the University has no financial stake in these companies, most of which are located next to its campus. Many of the companies retain close links with the University but their staff have taken the decision to leave academic life. None are employed by the University.

Chalmers University of Technology in Gothenburg has been the origin for more than 150 new companies since the early 1970s and it is claimed that as a group these companies are now the largest technical employer in the region. A chair of innovation was established at Chalmers in 1983 (TII, 1990).

The University of Twente in the Netherlands set up the TOP programme in 1984 to encourage its graduates to start their own companies. Under this scheme the graduates receive payment to work part-time in a university research team and are free to spend the remainder of their time on their own business ideas. TOP graduates are supported by experienced entrepreneurs and provided with industrial contacts through TRD, the university's technology-transfer research & development group. A total of 94 new firms had been created by 1992 and only 16 had failed. Of the remainder, 30 were in production of various kinds and 48 were in consultancy. TOP and a similar scheme with the acronym TOS, which aims to develop new product ideas, have had considerable impact on the regional economy. The companies have directly created 445 new jobs and assuming a multiplier of six new jobs in supplier companies, it is believed that 2,500 jobs have been created at very low cost (Kobus, 1992).

The University of Liege has formed 25 spin-out companies between 1986 and 1994 in biotechnology, mechanical engineering and medicine. The University has invested in 14 of these and has supplied space and consultancy to the remainder. Turnover is ECU 38m, 250 jobs have been created to 1994 and the aim is to create four new enterprises each year. A Science Park has been set up in conjunction with universities of Aachen and Maastricht and these three universities had a total of 70 spin-offs in 1993.

USA

In Europe, few established academics (as distinct from recently qualified post-doctorates for example) have actually left university employment to become entrepreneurs. In the USA, by contrast, there is quite a long tradition of academics giving up university life to form companies, often with the encouragement of their universities, for example see Bagalay (1992). Many former academics have founded companies which had great impact on their local economies. Some, such as Frederick Olsen who left MIT in 1957 to form what became Digital Equipment Company, started companies which became leaders in their field. DEC's revenues in 1990 were in the region of $30 billion. Roberts describes many other examples of entrepreneurial activity arising from MIT, Stanford and elsewhere in USA. He believes that 'positive-feedback' from earlier successes stimulates and encourages subsequent entrepreneurship (Roberts, ibid.).

Speaking at a DTI sponsored conference in London in 1992, the Director of MIT's Technology Licensing Office, John Preston, quoted a recent study by the Bank of Boston which had identified 636 spin-outs from MIT located in Massachusetts alone.

These companies had 1988 revenues of $39.7 billion. To put these revenues into perspective, the State Gross Product of Massachusetts in 1988 was about $120 billion. In addition, economists tell us that for every high-technology job created, four or five low-technology jobs will be created (Preston, 1992a).

MIT currently spins-out between four and ten high-technology-based companies each year plus five or six of what Lita Nelsen of the MIT Technology Transfer Office refers to as 'hobby companies'. It is a measure of the ubiquity of these firms and the particular ethos which exists at MIT that they only begin to count their spin-out companies when their annual turnover exceeds $500k.

Although something of a pioneer and probably the best known American university in this context, MIT is not alone in spawning technology-based companies. One of the early MIT researchers who moved to Stanford, Frederick Terman, was instrumental in helping, amongst many less famous entrepreneurs, William Hewlett and David Packard. The University of California and many other American universities now have impressive records and for every Hewlett Packard or DEC there are hundreds of lesser known successful companies.

Drawing mainly on American experience but with many similarities to experience in Europe, Etzkowitz (1990) described what he referred to as a second academic revolution in which economic development is added to the traditional roles of teaching and research as a legitimate function of a university. This process as described by Etzkowitz is an evolutionary one in which many academic research groups have grown to become quasi-firms, then in some cases to be formed into actual firms, often with the initial objective of raising research funds. Universities have also been faced, through financial stringency, with a necessity to capitalise their research, through licensing and technology-transfer to industry. This led in due course to incubator units for new products and the agglomeration, adjacent to the university, of commercial research based activity and the development of science parks. Beyond this, universities have adopted the role of venture capitalists in support of technology-based economic development. This process was in practice much less linear and uninterrupted than as described by Etzkowitz. Nevertheless the increasing contribution of universities to economic development is evident not only in USA but in Britain and the more advanced countries of Europe.

Taxonomy of University Companies

Lowe's paper provides a taxonomy, with examples, of university-linked enterprises. These range from in-house research centres, which are completely within the university's control, to spin-out companies which have been formed by current or former members of academic staff or which use university intellectual property but in which the institution has no direct financial stake or control. The exploitation structures were also categorised by function, in a spectrum extending from flexible activities such as consultancy and contract research through progressively more specific activities such as testing and collaborative research to prototype manufacture and routine production (Lowe, ibid., Chapters 9-10).

This categorisation follows the transition from 'soft' to 'hard' companies described by Matthew Bullock (1983) who pointed out that, typically, soft companies begin as consultancies selling the expertise of one or more academics to solve a particular client's problems. Gradually there follows a 'hardening' process during which the products develop to become more generally applicable and less client specific. This process may continue through, for example, the provision of design and testing services and in some cases to the extent where manufacturing begins. Bullock (op. cit.) lists some of the features of a typical soft company:

- The gradual hardening of the business from selling high level consultancy to the manufacture of a standardised product;
- The tendency for work to be undertaken on behalf of a specific client on a bespoke contractual basis, rather being produced on spec for a general and relatively inexpert market;

- The overlapping of subsequent stages in the hardening process so as to ensure a balanced portfolio of sales and cash flow;
- Concentration on a fairly narrow range of customers in technical, expert markets, primarily in government and industry;
- Tendency for the entrepreneurs to on-sell the company to one of its major customers rather than trying to grow it into a large company.

This pattern is recognisable in Lowe's study, in the development of the Swedish spin-out companies and in many of the UK university companies surveyed during the current research. Roberts' research at MIT however shows that only a few of those firms which began as consultancies or contract research companies have become high performers, in contrast to those which are product-oriented.

> ... product orientation alone does not ensure success. But lack of such orientation almost certainly assures lack of high growth and high profitability (Roberts, ibid.).

The marketing and delivery of university services and facilities, including staff consultancy, a role more usually carried out by university industrial liaison offices, is now provided by some university-owned companies. One of the first such companies to be formed was AURIS Limited, at the University of Aberdeen. AURIS has a number of trading divisions. Similarly Impel markets the consultancy services of Imperial College, London and ULIS Limited those of the University of Leeds.

Although various authors have made a clear distinction between university-owned companies and other spin-out companies there is no universally agreed terminology, either in the UK or internationally. A discussion paper issued by the Irish National Board for Science and Technology in 1987, for example, defined a 'spin-off' company as one in which, inter-alia,

> ... the founding entrepreneurs have worked, immediately before incorporating the company, for a university or a larger company generally in the same industrial sector (NBST, 1987a).

Studies of spin-outs from MIT, excluded

... situations in which the ex-MIT lab employee, even though a key person in the new company, was not actually a founder of it. ...(and) all situations in which the company, although new in terms of organisational form, was created as part of or as a subsidiary of another firm already in existence (Roberts, 1991).

The criterion for inclusion in a Swedish study by Olofsson & Wahlbin (1984) was that at least one founder was employed at the university when the company was founded and that the business idea was to commercialise knowledge and technology developed at the university. Samsom and Gurdon (1990) in a study of 22 recently started biomedical enterprises in Canada and the USA, included those firms which had been founded by a scientist-entrepreneur

... whose primary occupation, prior to playing a role in the venture start-up, and possibly concurrent with that process, was that of a clinician, lecturer or teacher affiliated with a university, research institution and/or hospital.

Referring to American experience, but in a summary that could be applied more universally, Etzkowitz (1983) described university companies and their products

Some of these firms produce, in quantity, particular products such as industrial enzymes or integrated circuits. There are private firms, owned by academic scientists, which produce substances or devices such as genetically engineered bacteria or new methods for locating mineral deposits, which are sold to other firms. There are also firms which produce small quantities of 'Quality Material' for use in research such as new cell lines or prototypes of integrated circuits. Finally, there are consulting firms that produce reports, market surveys recommendations and other proprietary information. A single firm may engage in more than one of these activities. What is distinctive about these firms is that they have been initiated with the active participation of scientists who hold academic appointments while they are participating in them.

Summary

Although university spin-out companies have long been a feature of academic life in the USA they are a comparatively recent addition to technology-transfer mechanisms in Europe. Since the early 1980s however there has been a rapid increase in the number of companies formed either by universities or by academic entrepreneurs, and sometimes by both, to exploit commercially university originated research. Unlike the American experience - where university spin-out firms have made considerable economic impact - it is regrettable that in Europe, although individual firms have been quite successful by most criteria, none has yet grown to become a major influence on its sector.

A taxonomy, based on work by Bullock, suggests that university companies lie somewhere on a spectrum ranging from 'soft' companies which are typically selling consultancy, to 'hard' companies which manufacture products. There is however no widely accepted definition of these companies which are variously referred to as spin-outs, spin-offs, campus companies and university firms (see note 1 in Chapter 1).

7 Critical Issues

Introduction

The literature on university companies has highlighted a number of critical issues affecting firstly, the decision to form a university company and secondly, the prospects for its success. These are not of course the only considerations in company formation and operation. A great many factors including the likely market demand for the proposed company's products or services, the availability of funding, of staffing and of physical resources such as buildings and space, are also important. These latter factors however apply to the formation of any new enterprise, are not unique to university companies and are not considered here.

Why University Companies are Formed

There are a great many references in the literature to explain the motivations and influences which affect the formation of technology-based firms and university spin-outs. Most of these however concentrate of the motivations of individuals, often of academics. There is relatively little to explain the motivations of the universities which engage in entrepreneurial activity. In a literature review Cooper (1986) classified six environmental factors, some of which could apply to an institution but most of which are more relevant to the individual. These are economic conditions, access to venture capital, examples of previous entrepreneurship, opportunities for interim consulting, availability of support personnel and services and access to customers. The received wisdom in most of academia and the most frequently offered reason by universities to explain their venture into new company formation is to provide an additional source of revenue in an increasing era of underfunding (Etzkowitz, 1990). That however is almost certainly too simple an explanation, perhaps even a rationalisation and there is evidence that there have been other motivations, for example, a desire to play a role in local economic development see Beveridge (1991).

It would have been surprising if the emergence of the enterprise culture of the 1980s had not spilled over into the universities. The growth of the science park movement[1] to encourage the clustering of high-technology firms was also a stimulus and even universities which were not directly associated with science parks began to form companies, (Monck et al., 1990). Initially, the initiative for new company formation often came from individual academics and their institutions responded with varying degrees of enthusiasm. Although, in hindsight, a great deal of the publicity received by the pioneers was less than objective, and by no means all of the new companies were successful, the movement gained momentum so that it seemed that for a university not to have at least one company was to be left behind.

An important factor influencing developments in the UK, is that intellectual property rights (IPR) arising from academic research, whether patentable or not, belong not to the researchers but to the universities which employ them. This is not the case in USA or in many European countries, where ownership is vested in the researchers or inventors themselves. Nevertheless most UK universities have arrangements to share the proceeds of exploiting their IPR with the staff members concerned. Individual researchers are, by law, named as the inventors of university-owned patents. Some institutions actively encourage their academic staff to bring forward their ideas for evaluation and indeed the first moves to establish many of the university-owned companies in the UK came from the engineers and scientists themselves. In a different legislative climate these scientists and engineers might well have formed companies on their own as, for example, did most of the spin-outs from the University of Linköping in Sweden.

One effect of this legislation is that, although academic staff are frequently involved as shareholders or directors and in some cases also in executive roles in companies formed by British universities, the primary responsibility to exploit the technology rests with the institutions. Universities in the UK therefore have a greater motivation to start companies themselves and retain an interest and involvement in them than is the case elsewhere.

Motivations of Researchers

There is little doubt that the attitudes of their peers, as well as those of the universities, are significant factors in encouraging or deterring academic entrepreneurs. So too are their personal motivations and expectations. Aspirations affect attitudes and determination to succeed as well as perceptions of what constitutes success. The motivations of key researchers may or may not be compatible with those of their universities or of others who may be required to provide funding. Even where the aims of the researchers and their institutions can be harmonised, their separate expectations will affect the kind of support structures and framework needed if there is to be any hope of achieving them.

According to Samsom and Gurdon (1990) the most prevalent perspectives relate to seeing the enterprise as an opportunity to advance scientific objectives and vision and to build a successful business as a matter of personal achievement. Smilor et al. (1990) distinguished between push factors, which were not found to be of great significance, and pull factors which were the dominant reason given for academic spin-outs from the University of Texas at Austin.

The most significant push factor was the need for additional money. Others included lack of excitement with a university career, difficulties in dealing with university bureaucracy, general frustration in dealing with the university, dislike of research responsibilities and requirements and a rejection of ideas within the university. The primary pull influences were recognition of a market opportunity, desire to try something new, desire to put theory into practice, the prospect of business contracts, the desire to start a company and to have fun with an entrepreneurial venture. The authors concluded that entrepreneurs from the University of Texas at Austin developed spin-out companies not because of dissatisfaction with the university environment but because of other motivations. This finding agrees with that of Segal Quince and Wicksteed (1990) who estimated about 70 per cent of entrepreneurs in 261 NTBFs in Cambridge stated pull factors as being their most important motivating factors.

In a comparative study of NTBFs in the UK (Cambridge) and Finland, Lumme et al. (1992) found strong encouragement in both countries for academics to go into business. Many however only did so because they found it to be personally interesting. The motivations of the founders were mainly technologically oriented and negative push factors were not important as motives. Amongst the top six reasons for the decision to form companies in both the Cambridge and the Finnish samples were the following:

- Desire for developing one's own ideas;
- Top-class technological know-how;
- Own professional skills, talents and experience in the field;
- Good marketing opportunities of the product;
- Identified customer needs or identified defects in existing products.

By contrast, Weatherston (1993 and 1994) concluded that, in the UK, push factors were high and negative influences were a factor in the decision by academic entrepreneurs to spin-out companies and that they are no different in this respect from other entrepreneurs. Weatherson's study in 1992 which involved 26 academic entrepreneurs from 12 universities showed that the majority of academic entrepreneurs maintain close links with their institution and spend only a small proportion of time working in the company. Most did not want to enter business life full-time, and this, it is concluded, is sure to have an adverse effect on company growth. Weatherston also concludes that academics are not being offered the opportunity to be enterprising and that university policy makers need to be aware that academics are being pushed into spinning-off companies and it may be necessary for universities to adopt more supportive policies. This is a more controversial finding and although it has been found that amongst entrepreneurs in Canada limited career prospects were often the impetus for academics to start a business (Doutriaux, 1987), this was not apparent in a more recent study of academic entrepreneurs in the UK (Harvey, 1994).

These are important issues. An understanding of the motivations of entrepreneurial universities and their researchers matters not only to those directly involved but should be of concern to others, such as government development agencies that wish to encourage the exploitation of research to benefit the local or national economy.

Possible Conflicts of Interest

Differing individual and corporate motivations raise the possibility of conflicts of interest. For example, the traditional academic reward system based on publications may well be in conflict with the need to maintain commercial secrecy in order to protect intellectual property.

There are also difficulties in attempts to reconcile traditional academic attitudes to research with needs for external funds and also opportunities for researchers to benefit personally, i.e. financially, from exploiting their research. Etzkowitz (1983) noted the uneasiness about potential conflicts of interest by the founders of many firms who had maintained their academic positions. To quote Etzkowitz

> ... Scientists see it (the commercial utilisation of their research) as a source of income for themselves as well as for ampler support for their research; academic administrators see it as a means of support for universities; businessmen see it as a source of profit, and governmental officials as a means to revive the economies of cities, states and of the country as a whole.

Samsom and Gurdon (ibid.) also noted problems experienced by scientists in coping with their implied duality in loyalties in working for a greater cause than just the organisational mission. In reality the aspirations and expectations of those involved in the formation of university-based companies, which may include university departments and administrations and sometimes industrial collaborators as well as researchers, are frequently different. Provided they are not mutually exclusive, however, differing objectives need not be a problem.

A more recent tension has been caused for universities in the UK. On the one hand government is urging universities to undertake industrially relevant and exploitable research, while on the other, the funding formula introduced by the Higher Education Funding Councils determines research income to universities on the results of periodic assessment of research quality. Since these quality assessments are based primarily on peer-reviewed publications, the pressure to obtain or maintain research ratings may well deter academics from engaging in what are perceived to be less prestigious commercial activities i.e. the pressure to publish may well be detrimental to commercial exploitation of university research.

A more legalistic matter is raised in cases where an official or employee of the institution also acts as a director of a company. He/she will owe separate duties of care to the institution and to the company to act at all times in the best interests of each (Robson Rhodes, 1996).

Academic to Entrepreneur - Difficulties of Transition

The transition from academic researcher to new company entrepreneur has not always been an easy one to make. Until comparatively recently it may have been easier in the USA than in Europe, although, as Matthew Bullock has observed, most academics start down the road to commercialisation with misgivings and hesitancy. American universities have lowered the threshold to commercialisation, mainly through their attitude to the involvement of academic staff in commercial activity. This has encouraged the emergence of what Bullock referred to as 'soft companies' (Bullock, ibid.). These it is argued provide an easy entry for academics to the commercial world but as has been shown, by Roberts and others, such companies rarely achieve the growth rates or financial success of more product oriented companies.

Louis et al. (ibid.) in a study of Life Scientists located in major American research universities, defined five basic forms of entrepreneurship and hypothesised that individual characteristics are moderated by institutional location, and they offer some possible explanations:

- Self selection - in that individuals are drawn to those institutions which are known to be supportive of or to tolerate entrepreneurship;
- Individuals are influenced by the behaviour of their immediate colleagues and tend to act like them;
- Organisational culture may be a factor (a broader set of organisational policies, procedures and values reinforces attitudes and behaviour regarding entrepreneurship);
- Strategic management may be a factor (Some universities use recruitment to position themselves in the forefront of changing patterns of academic behaviour to reap the potential benefits in increased prestige and income).

Even in America, where academic entrepreneurship on a significant scale predated that in Europe, many who have attempted to combine their business and academic activities have met with adverse reactions from erstwhile academic colleagues (Etzkowitz, 1983).

Some of the criticism comes from those who believe that the proper role of the universities is to carry out basic, curiosity-driven research and that it is inappropriate to be diverted into commercialisation. They have a case. One can argue that without basic research to open up new and unexpected possibilities, all that can be achieved are incremental improvements within the boundaries of existing knowledge. In hindsight it is easy to identify discoveries arising from 'pure' research which enabled great advances to be made. Understanding of the DNA helix, the transistor and penicillin for example, did not come from applied research or attempts to commercialise biochemistry, physics or chemistry, yet without these seminal discoveries today's biotechnology industry, the information technology revolution or the subsequent developments in antibiotics would not have been possible.

Despite these views it is becoming more 'respectable', at least in the UK, for academic researchers to undertake commercial activity. This is a fairly recent phenomenon which may be due to changing perceptions about appropriate activities for a university. It may be a sign of acceptance of the new reality that an increasing proportion of university research funding must come from the private sector.

The Role of the Entrepreneur

Despite there being no universally agreed meaning of the term, it is conventional wisdom that an entrepreneur is an essential ingredient for a successful new company start-up. Although this requirement is not unique to university spin-outs it is of particular importance that it should be consciously considered in a culture not normally accustomed to or noted for entrepreneurship. The role of entrepreneurs, who are usually understood to possess some combination of business and technical acumen, and of so-called 'product champions'[2] at various stages of company growth, bears closer examination. In a painstaking historical review, Hébert and Link (1988), traced the changing role of the entrepreneur from the eighteenth century to modern times. They summarise their findings by a variety of concepts of the role of the entrepreneur, who is:

- The person who assumes the risk associated with uncertainty;
- A supplier of financial capital;
- An innovator;
- A decision maker;
- An industrial leader;
- A manager or superintendent;
- An organiser or coordinator of economic resources;
- A proprietor of an enterprise;
- An employer of factors of production;
- A contractor;
- An arbitrageur;
- The person who allocates resources to alternative uses.

Some of these concepts overlap but all past theories of entrepreneurship have focussed on uncertainty and innovation or on some combination of the two. Hébert and Link make the point that uncertainty is a consequence of change whereas innovation is primarily a cause of change.

> Thus, if we accept the fact of change, we must therefore expect entrepreneurship to have two faces - one that reveals itself when the level of inquiry deals with an explanation of change, the other when investigation concerns itself with the effects of change (Hébert and Link, 1988).

Stankiewicz (1986) offers the following definition

> An entrepreneur is an agent (individual or collective) who champions a novel combination of productive resources as the means of achieving an economic end. He brings about, in other words, a reallocation and recombination of the factors of production. The entrepreneur articulates the commercial goal, identifies and solicits the necessary resources, coordinates their use, and assumes the ultimate responsibility for the success of the project. He is the very opposite of a specialised technician or a conventional manager; his role has no clear boundaries but instead expands and contracts depending on the character and the stage of development of the enterprise.

Oakey et al. (1988) note the need for both technical ability and business acumen which can be present in varying degrees and combinations - the authors offer an entrepreneurial matrix on which individuals can be placed.

> The entrepreneur is most commonly defined as an individual who, through his personal drive and novel technical or managerial insights, is able to achieve significant growth for the firm which he either owns or manages for other providers of capital. In common parlance, an individual does not qualify as an entrepreneur until he has become successful; and because the concept of a charismatic leader, sweeping all before him, is both romantic and simple to understand, the entrepreneur has become a major plank of much right-wing economic and political philosophy in recent years. Moreover, since entrepreneurs tend, by definition, to be individuals, there is great potential for the creation of personality cults where entrepreneurship is not a 'cold' concept, but can be exemplified through live examples (Oakey, Rothwell and Cooper, 1988).

The most successful entrepreneurs, particularly in the context of technology-based firms, are likely to be those who have acquired both technical skills and business acumen. The need for business training for scientist-entrepreneurs was highlighted by Samsom and Gurdon (1990). In a study of recently started Canadian and American biomedical firms, half of the scientists discussed one or more specific functional skills which, after some time in the venture, they found themselves to be lacking. Most significant amongst these were team management and interpersonal skills followed by financial and marketing skills. Two-thirds of the scientists recognised a need for business training but none acted on this need.

A similar situation was reported in a study of 25 academic entrepreneurs from UK universities by Harvey (1994). Only three of these (12 per cent) had any prior business experience, four (16 per cent) had prior similar work experience and 18 (72 per cent) were complete novices with regard to business experience. This characteristic, claims Harvey, is shared with US academic entrepreneurs. Perhaps it is not surprising that academics should have little business experience but it is regrettable that none of the 25 in the sample studied made any attempt to participate in business or entrepreneurial training. Of course it is a moot point whether entrepreneurship can be taught. There are those who claim that it can (e.g. Bygrave, 1995) and those who assert equally passionately that it can not, although it is conceded that the skills necessary to sustain successful entrepreneurial activity can (e.g. Kinsella, 1995a). This is an ongoing debate in which the protagonists tend to hold very strong opinions.

In the study by Rothwell and Dodgson (ibid.) for IRDAC the greatest concern about future obstacles to growth related to shortages of key personnel, including lack of entrepreneurial management, six firms.

Despite the growing consensus about the importance of entrepreneurship, its usefulness as a concept to the growth of NTBFs and the variable impact of entrepreneurs over time has been questioned. Oakey et al. (ibid.) note the difficulties represented by:

- Plural entrepreneurship in which entrepreneurial acts are frequently those of a small number of individual founders acting in concert;
- The variable impact of entrepreneurs on a company over time. There is evidence that the single entrepreneur founding and steering a company to ultimate success is over-simplistic and atypical;
- The dubious advantage of entrepreneurship to high-technology small firm growth. Many of the so-called advantages of the entrepreneur may be inappropriate, especially after the initial embryonic phase. Characteristics such as an unwillingness to delegate and to share decision making may inhibit company growth. Once a company has passed its initial stages it may be argued that the company would be better served by professional managers rather than being under the control of the original entrepreneur.

Nevertheless, the authors acknowledge the vital contribution of entrepreneurs, either singly or in concert, to the birth and growth of new high-technology-based companies.

Roberts (1991) identified five different key roles which must be fulfilled in any technology-based firm. These are filled by:

- Creative scientists who are the source of new technological innovations;
- Entrepreneurs or product champions who push the technical idea forward in the organisation;
- Project managers who focus on the specifics of the new development, decide which aspects warrant further development, and coordinate the needed efforts;

- Sponsors who protect the scientists and entrepreneurs so that innovative technical ideas survive past the initial development stage;
- Gate-keepers who bring essential technical and market information into the technical organisation.

In research spanning 25 years, Roberts discovered a strong correlation between success of technology-based firms and the number of original company founders. For example, 14 per cent of multiple founder companies from MIT labs were in the highest performance group, while only three per cent of single founder enterprises attained that level. This was explained as being due to the greater variety and depth of talents, capabilities, initial capital and experience available to new firms as the number of founders increases.

It may of course be argued that not all founders are necessarily entrepreneurs, as the term is now becoming understood. Some may simply be financiers who take no active part other than to monitor their investment. The benefits of complementary skills is recognised by Soderstrom, Carpenter and Postma (1986) in an account of practices adopted at Oak Ridge National Laboratory (ORNL) under contract with US Department of Energy. They point out that - if innovations are to be generated, enhanced and moved forward in the organisation, a variety of individuals must also be involved. They believe that entrepreneurs are needed to complement the scientist and recognise the commercial potential and note that the ability to do this comes from a broad range of interests and also that most technology occurs from one to one interactions.

Roberts (ibid.) has found that family background, age and experience, personality and motivation all affect the development of a technical entrepreneur although family background variables show no direct effect on company success. Roberts concludes that entrepreneurs are made and not born. Other characteristics which have been identified by various researchers include, gender, sibling position, family involvement in small business, exposure to entrepreneurial role model, age on founding a first business, number of businesses founded and business experience. Harvey (1994) found good agreement with results in a sample of UK academic entrepreneurs and research reported from USA.

In a comparative study of entrepreneurship in Northern Ireland and the Republic of Ireland it was found that entrepreneurs who establish more than one business tend to concentrate their activities in one sector rather than across all forms of possible enterprise. Experience, knowledge and expertise developed in one sector tend not to be applied to another, Pickles and O'Farrell (1987).

The External Environment

The general economic climate as well as government policies affect both attitudes to new company formation and their subsequent performance. Oakey (1991b) argued that the British Government failed adequately to stimulate and ensure the growth of NTBFs during the previous decade. Certainly influences in the local or regional environment including the attitudes of local government development agencies can have a profound affect. It has been argued that in regions less well endowed with technology-based firms the local universities have an important role to play as seedbeds for entrepreneurship. Where this is recognised by local government development agencies to the extent of providing encouragement and financial support for university spin-outs, the prospects for their success will be enhanced. Widespread local support for commercial exploitation of university research through entrepreneurship also engenders a degree of approval and legitimisation of this type of activity (NIEC, 1993, para., 8.12).

Summary

Universities have formed companies for a variety of reasons. The most commonly expressed reason is to obtain funding to supplement that available from Government but it is by no means clear that this is the only motivation. Since UK universities usually own the intellectual property arising from their research the onus is on the institution as distinct from its academic staff to take the initiative to exploit it.

The motivation of individual academics to become entrepreneurs is even more complex. Most research has found that pull factors are more important as a spur than push factors. The former include the desire to advance scientific objectives, to build a successful business as a matter of personal achievement, the desire to put theory into practice and the opportunity to see their ideas developed and applied commercially.

Conflicts of interest are to be expected between the university and commercial roles of academics and members of university administrations involved in company activities. Tensions are also expected to occur as a result of apparent contradictions between the UK Government's exhortations to universities to become more industrially relevant and Government funding for research which is based on assessments of research policy as determined largely by peer-reviewed publications.

The presence of one or more entrepreneurs is widely regarded as being essential to ensure the success of a university company. Although most researchers agree that the role of the entrepreneur is focussed on dealing with risk and uncertainty, there are differences of interpretation about the precise role of an entrepreneur in these circumstances. Those most likely to succeed are likely to have acquired both technical skills and business acumen. There is a need for complementary skills and it has been noted that the original founders are not necessarily the best placed to fulfil the entrepreneurial role.

Closely related to difficulties arising from conflicts of interest are problems of transition from academic to new company entrepreneur. These are due in part to individual characteristics but are also affected by the institutional location and ethos. Criticism from erstwhile colleagues has occurred, some of it from those who believe that the role of researchers is to create new knowledge without becoming involved in its commercialisation. Recent evidence suggests however that it is becoming more 'respectable' for academics to be involved in this type of activity.

The general economic climate as well as government policies plays a part both in influencing the formation of university companies and in their subsequent success One would reasonably expect therefore that these factors as well as regional considerations would be taken into account before spin-out firms were founded and that their affects on performance would subsequently be identifiable.

Notes

1. The growth and effects of the Science Park movement has not been considered in detail here. For an overview see the following: Oakey, 1985; Monck et al., 1990; Industry and Higher Education 9 (6) 1995; Porter et al.,

1990; Westhead and Storey, 1994; Gemunden and Heydebreck, 1994; The Times, 1994.

2. Typically product champions are individuals, often inventors, with an unshakeable and occasionally unrealistic faith in the value of the product being developed. They tend to be characterised by a tenacious desire to see the product succeed and frequently make great efforts to overcome set-backs. In the context of university spin-outs they are often the most technologically aware of the participants and may be frustrated by the relative importance accorded to other more commercial considerations

8 Towards a Better Understanding of University Companies

Introduction

Earlier reference was made to the fact that although a great deal of research has appeared in the academic literature about university companies and about NTBFs in general, relatively little had been recorded about university spin-outs by practitioners in the universities. In this Chapter the experiences and insights of this group are reported and discussed for comparison and contrast with the published literature. In particular the relevance and importance of the critical issues described in the previous Chapter will be tested in the light of recent experiences. The interview programme to be described also provided an opportunity to explore similarities and differences amongst universities and their spin-out firms.

The Interview Programme

The first tranche of interviews, which were recorded verbatim and later transcribed, were held at 16 universities in England, Scotland, Northern Ireland and the Republic of Ireland which, at the time of the interviews had spawned a total of more than 200 new technology-based firms (NTBFs). To obtain additional information about university holding companies - which emerged as a growing form of university company structure - separate visits were made to five universities, four in UK and one in the Republic of Ireland. For overseas comparisons two universities in Sweden were visited, one in the Netherlands and three in USA. Additional material on technology-transfer in general was obtained from interviews with members of staff at British Technology Group (BTG) in London and staff at the US National Technology Transfer Centre in Wheeling West Virginia. A complete list of the institutions visited is given in the Appendix. The information was given in confidence so individual institutions are not identified in the text, except where permission has been obtained.

As there is no generally agreed definition for a spin-out company, it was decided to confine the initial interviews to universities which had set up companies to exploit their technology or expertise and which maintained a financial interest in at least some of their spin-outs. Later information was gathered, for comparison and contrast, from university spin-outs in which the founding university had no financial stake.

Most of those interviewed were university staff members who were administratively responsible for the companies either currently or at the time of their formation and able to speak knowledgeably about their origins and development. A number of the interviewees had personally been responsible for or significantly influential in determining their university's policy and procedures for establishing companies and some were shareholders or directors of spin-out companies.

The Interviews

The questionnaire was designed to be administered during face-to-face meetings lasting between 45 and 90 minutes Not all questions were relevant to all universities and in some cases questions were omitted; others were modified or expressed differently to suit the circumstances. Supplementary questions were asked in response to many of the answers. Within the constraints of time and maintaining the structure of the meetings, all the interviewees were encouraged to talk freely and to amplify their replies with personal opinions. An undertaking to respect confidences and preserve anonymity no doubt contributed to the candour with which most questions were answered. The interviews were recorded verbatim and subsequently transcribed. In the text which follows many of the responses are quoted at length, not only for their content but to try and convey something of the 'tone' and 'flavour' of the replies received.

The Topics Explored

The questionnaire was designed to obtain the views of practitioners about those issues identified in Chapter 7 as well as to obtain a rounded overview of current practice. It was designed to capture information, both factual and subjective, under a number of broad headings.

Overview of Companies:

- The numbers and types of companies at each university;
- The businesses they were in and their approximate sales turnover;
- Their structure and how they were funded;
- The nature and extent of the university's investment, if any;
- The patterns which led to company formation;
- The relative influences of technology-push and market-pull;
- The existence of entrepreneurs or product champions;
- Views on how successful the companies have been;
- What was judged to constitute 'success' and how this was assessed;
- Whether alternatives to company formation were considered.

The University's Attitudes and Structures:

- The nature and extent of university involvement;
- The university's attitude to company formation;
- The university's procedure, if any, for company formation;
- Whether there were systematic means of identifying commercially relevant research;
- The formal and informal links between the university and the companies.

University Staff Involvement:

- The extent and nature of university staff involvement;
- The motivations of academic staff who became founders of companies;
- The motivations and roles of other university staff to become involved;
- Conflicts of interest, if any, between company and university roles.

104 *Campus Companies - UK and Ireland*

Effect of the External Environment:

- External factors which affected the decision to form the companies;
- Factors which either helped or hindered the companies' operations or growth;
- The effect of the local or regional economy on operations;
- Effects of other universities, proximity of Science Parks or similar;
- Government financial pressure on universities, scarcity of research funding.

Looking Ahead:

- Future plans;
- Changes in the university's role;
- What growth was forecast;
- Limiting factors to growth;
- Additional investment needed and its source.

Looking Back with Benefit of Hindsight:

- The original expectations of the founders and the extent to which they had been fulfilled;
- The factors which had been most influential in shaping the companies;
- The ranked importance of a list of factors to the success of the companies;
- The single greatest mistake or omission;
- Any other lessons learned.

Issues Arising From Interviews

The responses of the interviewees to some of the questions turned out to be much more revealing than others and provided insight into the similarities and differences at the various universities. A number of matters were of widespread concern. Others were more significant at some universities and of less importance elsewhere and consequently were dealt with in different ways.

The main issues addressed in this Chapter are:

- Why companies are formed; how decisions to form companies are taken; the effect of external factors on company formation and operation; how company success or performance is judged;
- Motivations of would -be academic 'entrepreneurs';
- Conflicts of interest for university staff;
- The relative importance of market-pull and technology-push;
- The importance of entrepreneurs and product champions.

Additionally, because of the understanding they provided, details are included of the responses to questions about the extent to which, in hindsight, original expectations had been met and about the relative importance of various factors to the success of companies

Reasons for Company Formation

Why Select the Company Route to Exploitation?

The intention here was to explore the reasons why universities formed companies in preference to other means of exploiting their research and intellectual property. It is important to remember that most of those interviewed were senior university administrators - typically industrial liaison directors, industrial unit managers, or managers of university holding companies - who were giving their views about the motivations of the university as the founder of the companies and who were well placed to reflect that view.

Only 60 per cent of universities claimed to have seriously evaluated alternative means of exploiting their technology before deciding to form new companies. Various reasons were given for this, but many of the replies suggested that the pressure to form companies had come initially not from the university but from enthusiastic academics, three examples follow.

'In all of the cases I've dealt with here, its a second career move for the academic in question. They tend to be mature academics who have reached a certain level of status in their profession, they have got into some niche. Very often they are people who are not going to be heads of departments or professors in this University and they are certainly not going to leave this University because it is a nice place to be. It is a good place to work and they are looking for another element of self-expression which they don't get in their present work.'

'They are always associated with people, heads of department or senior professors who have a wider knowledge of perhaps the market than most, although that's not very much, and have a certain entrepreneurial (flair?)'

'I would say that people, as a generalisation, wanted to exploit technology which they had a feeling for and an adapted knowledge of but were prevented from fully doing so for some structural internal reasons in the organisations they were working for and at the same time saw a way, not of making megabucks out of it, but of making a living out of it.'

Another consideration was the availability of funding for research as it progresses from fundamental to near-market and the apparent preference of some funders, at the later stages, to deal with a company rather than a university.

'Sometimes because in a long-term natural progression of research from fundamental, strategic, applied, prototyping - that smooth sequence - they (the researchers) suddenly find that the funding rules turn upside down, that the Research Councils no longer will fund them and they turn to the more commercial sources of funding. And these people say, 'Oh you can have the money but you must form a company first' and there is a very abrupt change at that point in time. Sometimes the transition takes quite a long time for people to adjust to. People are saying, I'll form a company to get the money, and we (in the University administration) are saying well that's not really the best reason. And if it is really only if the academic staff through looking at it and discussing it feels that.. Yeah well a company would in fact be something I would be keen on, enthusiastic about...'

In this latter case, even though the university may have had reservations about the rationale for forming companies, it had at least recognised what was happening and appears to have remained in control of the decision whether to take the company route. It does however indicate another potential problem in that the views of researchers and those of their university about the reasons for company formation may well be at odds. If undetected this could cause difficulties later and in some cases this has led to different expectations of the companies by the university on the one hand and the academic entrepreneur on the other. This point is developed more fully below. Similar examples were quoted at other institutions and add weight to the critical importance of ensuring that the expectations of all the parties are understood and mutually compatible before company formation.

At another university, where academics have sought to form companies, alternative means of exploitation such as licensing and collaborative R&D have always been assessed case by case. The decision to form a company being taken only if that seemed, on balance, to be the most appropriate means of exploitation. The experience there has been that,

'... when you look back on them, certainly on the recent bunch, they have been formed when an external person has already parted with some money for some version of the product ... somebody has wanted a bit of it. In (named company) someone took a prototype before the company started (another company),...sold a bit of product, had orders for the product before we started the company.'

The observation that external investors or funders, usually an existing company, played a role in the university's decision to take the company route was repeated in slightly different ways by several of those interviewed. The important point seems to be that to get to the stage of persuading a potential investor to part with funds for a prototype product it is generally necessary to have something tangible - as opposed to simply an idea - to demonstrate. This implies that funds to produce the prototype have to be found by the fledgling company, either from its own resources or - as in some cases quoted - from Government schemes such as the, frequently referred to, DTI sponsored Small Firms Merit Awards for Research and Technology (SMART) award scheme.

Asked specifically which of a list of external influences had significantly affected the university's decision to form companies, the percentages of interviewees selecting each alternative were:

- Pressure from government   92 per cent;
- Scarcity of research funding   75 per cent;
- The proximity of a Science Park   58 per cent;
- Activities of other universities 42 per cent;
- Regional employment incentives   33 per cent;
- Government grants etc.   25 per cent.

In hindsight, an obvious omission from this list of reasons was 'pressure from researchers'. It was only at the analysis stage that the ubiquity of companies formed as a result of initiatives from academics became apparent.

It is clear that university companies have been formed for many different reasons and no discernible pattern emerged from the interviews. Implicit in a number of replies was that very often the academics had taken the initiative and that their universities had either acquiesced or been persuaded to become involved. It would be unwise to stifle enthusiasm, but further research could well show that some failures and disappointment might have been avoided and better returns obtained for the universities by transferring technology by other routes. For a discussion of alternatives to company formation see Olofsson and Wahlbin (1984).

It would be unfair to conclude that universities have been coerced into forming companies without due diligence but some would admit that, faced by quite senior academics who were enthusiastic to form companies, the alternatives were not given adequate consideration.

The Effect of External Factors on Company Formation and Operation

The purpose of these questions was to discover the extent to which factors outside the universities affected firstly their decisions to form companies and secondly, the operation of university companies. It was equally important to learn of external factors which influenced companies in a positive way and any which had a limiting or constraining effect. These environmental factors seem to be quite significant in some cases but were not always recognised.

There was great diversity in the responses but the four highest scoring factors were:

- Regional factors   45 per cent;
- University policies   36 per cent;
- Government encouragement/assistance   36 per cent;
- Lack of government funding for universities   27 per cent.

Regional factors Comments grouped into this category included local economic and demographic factors as well as local competition and the regional industrial milieu. The 'Cambridge phenomenon' and the tradition of entrepreneurship in the Edinburgh area were both quoted as positive factors. By contrast another university felt that its area was viewed unfavourably as a place for new businesses to locate compared to the M4 corridor. A congested town centre, high rents, and difficult planning permission were variously mentioned as inhibiting factors. In Northern Ireland the positive influence of having a ready supply of skilled labour contrasted with a university in the South East of England which quoted high labour costs and low unemployment as being a constraining factor. To quote,

'... the companies which have developed have been usually high-tech with small staffing and a lot of automation. That is the only way we could possibly think of starting a company in the area, so mainly it is advanced technology.'

There is nothing which a university can do about its geographical location, short of expanding or moving elsewhere, and the noted tendency of technology-based firms to cluster (Rothwell and Dodgson, ibid.) will always tend to make it easier for university companies in favoured areas to prosper. However as Vaessen and Wever (1993) have pointed out, since there is less competition in more 'hostile' areas new companies may actually grow more quickly. On the other hand that effect may be diluted because, as is confirmed in this research (see Chapter 15), technology-based firms are likely to be much more export oriented that other types of business and tend to compete in world markets as distinct from locally.

The availability of skilled staff and in particular a ready supply of graduates and those with higher degrees in appropriate disciplines may nevertheless provide an advantage in more peripheral areas. Furthermore, staff recruited are more likely to remain with their employer if there are fewer opportunities to move to competitors. This latter point was made very strongly by university companies in Northern Ireland who referred to the lower mobility of technical and scientific staff there than in other parts of the UK. It is doubtful however whether that effect is advantageous in the longer term. Most regions which are relatively bereft of technology-based firms would welcome the opportunity to expand their stock of innovative companies and as already noted (Chapter 3) amongst the most prolific sources of NTBFs are existing technology-based firms.

Government attitudes Positive factors reported included the proximity and support of Scottish Enterprise (Scotland) and strong support from government and a good grants regime (Northern Ireland). The positive attitude of government and its economic development agencies in Scotland and Northern Ireland contrasted with the complaint of the Director of an English Science Park that government provided little support. Paradoxically universities in the Republic of Ireland believed that cut-backs in Government funding had led to encouragement of entrepreneurial activity.

The combination of carrot and stick is a feature of the general climate in which the universities find themselves. Local development agencies have important roles to play in stimulating technology-transfer from the universities in their regions and closer liaison between the agencies and universities could only be beneficial. In the NI universities for example, representatives of local government agencies are invited to sit on university committees and research steering groups.

Reference has already been made (in Chapters 5 and 6) to the tensions caused by the contradictions between the way university research funds are allocated and government's exhortations to universities to exploit their technology commercially. One way to clarify the position would be for government to allocate funds separately for research, technology-transfer and entrepreneurial activity in proportion to its priorities. This would not be an easy exercise and whatever the outcome it would be controversial but by avoiding the issue government is forcing many universities to back away from worthwhile commercially relevant activities in favour of conducting peer-reviewed research.

There are already several successful programmes in the UK. About a quarter of universities, for example, referred to the DTI sponsored SMART award scheme as being particularly helpful in providing pump-priming funds for new product development. It is possible that more funds targetted in this way would be beneficial.

Another approach is advocated by Doutraiux (1991) who concluded, in a Canadian study, that government should sub-contract from NTBFs rather than support them directly. This is appealing but although new companies are free to bid for Government work it could be difficult for government to favour them overtly, given the EU regulations related to competitive tendering. It would also be worthwhile to study some of the schemes used overseas, in Denmark, Sweden, Finland and the TOP programme which has been very successful at University of Twente in the Netherlands for example. The opportunity to access European Union (EU) structural funds is also available in some regions of the UK, notably those designated as 'Objective 1' regions. At the time the survey was undertaken Objective 1 status in the UK was confined to Northern Ireland but this has since been extended to other regions including Liverpool and the Scottish Highlands and Islands. The significance of this for universities in these areas goes beyond that for their spin-out activities and extends to the availability of research funding more generally.

University policies University policies both overt and implicit often reflect the institution's interpretation of how to maximise its overall funding. This comment for example was typical of a number expressing similar views.

'The only constraint (on entrepreneurial activity) is the balance of what the University thinks would be prudent to spend between academia and more commercial activity. And I guess the pressure on that balance over the last couple of years has shifted in favour of academia because money's short.'

Attitudes to risk were also important. One university which had set up a holding company to manage its spin-outs was, in the opinion of its managing director, being denied opportunities for growth by fear on the part of the university of entering into joint ventures in which the holding company would have a minority stake. Similar timidity was echoed at another institution.

'This is a risk averting university so pressure to consider setting up companies must avoid diluting academic work. On the other hand companies may be able to generate funds to feed back into research. On the negative side there is the fear that the University's reputation could suffer if a company got into trouble and the University would have obligations etc.'

Many people felt that attitudes and perceptions of the role of universities and academics influenced companies and entrepreneurial activity. A mixed bag of responses included the following:

- The perception that universities should not be using publicly funded facilities for the personal advantage of staff;
- Capital providers and risk takers did not look to the universities;

- It has become respectable to be an academic entrepreneur '...there's nothing like emulation. If you see your buddy driving in and out every day in a Porsche, you are quite likely to say 'why not me too?'';
- One Scottish respondent felt that the Scots feel inhibited '... by their reluctance to adopt an attitude to ostentatious display of wealth as in England. There is not the same enticement to greed.';
- It is not generally appreciated that technology is a long-haul and as a result people are apt to become impatient.

Most of the interviews were undertaken between late 1990 and 1992, a period in which the UK economy was in recession and many of those interviewed were feeling its effects. Only one respondent however referred to the economy as a factor influencing company formation.

'Looking back on it I feel that we set up ... at a time when there was a little bit of growth in the economy, which just may have made life easier. Things are tightening up now (December 1990) and that means that research and development laboratories may have been buying more readily. For example, our partners (a major computing company) were able to push out and subcontract software to (the joint-venture Company) and support the R&D there, rather than write it in-house, more happily than they might have been in tighter times. For instance now they are battening-down the hatches. So we may have hit it on a growth phase of the economy, that may be a slight factor. There was a good grants regime around which again has been tightened up recently.'[1]

Other factors Other factors named which affected company formation and operation included:

- The proximity of science or technology parks. The influence of science or technology parks on the establishment of NTBFs has been questioned by a number of authors. For example, Quintas et al. (1992) suggest that although there are certainly worthwhile linkages between universities and local science parks in that company founders often come from the universities and that joint research projects occur which facilitate technology-transfer, the emergence of new university spin-outs in science parks is the exception rather than the rule;
- The provision of EC funding for research, which had changed attitudes to longer-term R&D projects. This has had both positive and negative

effects on new company formation. Competition for research funding under successive EU Research and Technological Development (RTD) Programmes has become progressively more competitive. Success rates for university applications range from one in ten to one in four with a mean of about one successful project in every five submitted. A frequent requirement is that the consortia applying for support should contain at least one company and there has been a tendency for companies to be favoured as lead partners. Some universities have sought to improve their success rate therefore by applying through university-linked companies. On the other hand the increase in longer term research funding from Europe - a typical project lasts three years and there are often opportunities for follow-up projects - has reduced one of the pressures acting on the universities to form companies.

- The ending of the monopoly of the British Technology Group (BTG) monopoly to exploit government funded R&D. As a consequence of legislation in 1950 (Treasury Circular 5/50), the National Research and Development Council (NRDC) was given first refusal to exploit all Research Council funded research in the UK. BTG which was formed in 1982, following a merger between NRDC and the National Enterprise Board (NEB), inherited this right. The effect was that the universities were obliged to show all potentially exploitable research first of all to BTG. This monopoly was rescinded in 1985, since when the universities have been free to make their own arrangements to exploit their intellectual property. In 1992 BTG was privatised. Funding came from a consortium of venture capitalists and a number of universities also acquired a shareholding. The British Technology Group Limited now competes with others to acquire university R&D for exploitation.

Despite the foregoing, it is doubtful whether the removal of the BTG monopoly had much effect on the rate of formation of university companies because most Research Council funding had traditionally been for basic as distinct from near market or applied research. Even today when universities are free to exploit their research wherever they can and although some Research Council funding is available for nearer to market activity, relatively little of this work has been suitable for exploitation. The reference to this factor during the survey was made by one university which had formed a company to market the ability of its researchers to carry out consultancy for industry. Much of the pool of expertise had been built up over many years and funded from a variety of sources, including the Research Councils. Before the ending of the monopoly, permission to exploit this know-how would have to have been obtained from BTG.

- The availability of joint venture partners. This reference was made by the relatively few, but growing, number of companies which are adopting the form of corporate venturing in which an established company is invited to take a financial stake in exchange for marketing and other commercial expertise. It is more fully developed in the case-study described in Chapter 12.
- The availability of 'recycled entrepreneurs' willing to start something new. This issue was raised by someone who - referring to the 'Cambridge Phenomenon'. Segal Quince and Wicksteed (1990) - observed that

> ... you get the phenomenon of recycled entrepreneurs of people who set up companies (who are) interested in repeating that process. We are starting to see a bit of that but not a lot as yet, maybe it's early days.

This may be more likely to occur if the entrepreneurs stick to one area of technology or one industrial sector because as Pickles and O'Farrell (1987) discovered, knowledge and expertise developed in one sector tend not to be applied to another.

- The university's availability of funds and its priorities for their disbursement. The funding of the companies in the survey ranged literally from a few pounds to many millions. Those at the lower end of the spectrum, companies requiring perhaps a few thousands of pounds, generally presented no problem to their sponsors or to their universities, whilst those at the other extreme either had been or were about to be publicly floated on the Stock Exchange. The difficulties seemed to arise for those in between. The sources of funding for these companies included, business angels i.e. private investors, venture capitalists, investments by researchers and university staff members, Government grants and competitive schemes - the SMART award scheme was widely used, corporate venturing i.e. investments by other companies. A few university firms were able to obtain limited funds to pump-prime new projects directly from their university or its holding company. Bank overdraft facilities were also used by established companies to cover temporary cash deficits.

Standeven (1993) shows how appropriate equity sources move from family and friends, through government sources, informal investors, venture capitalists, corporations to public issue as companies progress from very small and small through growth-oriented to super-growth prospects. It is important to emphasise that the funding requirements of technology-based firms go through several stages. Pre-start-up funds may be required to produce a prototype or to demonstrate the validity of an idea. Such sums are usually relatively modest but vitally necessary. This stage is followed by early money to establish a company, to provide necessary equipment or facilities as well as working capital until it begins to generate revenue from sales. Private sector money is almost always required and this early stage money is at high risk but the financial returns can also be high. As companies grow they become cash-hungry and although some growth may be funded from retained earnings there is almost always a need for second-round or development capital. These stages are similar to those described by Roberts (1991) as the start-up, initial growth and sustained growth stages.

In a study of the sources of venture capital in the UK Mason and Harrison (1994) note the lack of business angels when compared to the USA. They conclude that there is a need to raise awareness amongst wealthy individuals of informal business venture financing an investment, to more effectively mobilise and overcome inefficiencies in the informal venture capital market and they recommend that measures to make tax treatment of equity investment in unquoted companies no less advantageous than other forms of savings.

A pattern seems to be emerging in which banks prefer to borrow against collateral rather than engaging in formal technological assessments and since new companies are likely to have little in the way of tangible business assets, entrepreneurs are likely to find themselves constrained in the extent of debt they can borrow (Philpott, 1994) This is the stage at which government grants can be particularly useful. It is also the point at which university companies may have an advantage over their non-university equivalents by providing their embryonic offspring with access to university equipment, laboratories and other facilities at less than economic costs.

- The university's attitude and degree of support, especially from the Vice-Chancellor or Principal. A number of universities described how enthusiasm for company-formation and indeed for technology-transfer by any route had waxed and waned under successive vice-chancellors and principals. This is not surprising as the chief executive sets the tone in any organisation and people take their cue from the attitudes coming down from the top. It is worth noting however that in a few cases the entrepreneurial ethos had survived what were regarded as hostile regimes. In a literature review Lumme et al. (1992), while not denying the importance of support at the top, advocates a bottom-up approach and concludes that the fostering of entrepreneurship at universities should take place as much as possible at the departmental level but that middle level support is necessary to maintain an enterprise culture.
- Difficulties which universities have in competing in the market place. It was widely accepted by the practitioners as well as by the entrepreneurs interviewed that universities are weak at marketing. This finding is consistent with the findings of Lumme, (ibid); Rothwell and Dodgson (1987); Samsom and Gurdon (1988) and Cartin (1992). This is one of the reasons why the corporate venturing approach, in which an existing company with a market presence taking part of the equity and using its marketing muscle, has been so successful. Of all the factors affecting company performance marketing was scored highest (see Figure 8.1 and Table 8.1).

To summarise, although quite significant in some cases, environmental factors, were not always recognised as being of importance when the companies were formed. With hindsight though, all universities were able to identify a number of significant influences on their companies. However there was no pattern to the replies or lessons which could be applied universally. Nevertheless there is no doubt that environmental factors can have significant effects and should not be overlooked.

How Company Performance is Assessed

The responses to this question, which was intended to explore how the universities judged the success of their companies, actually revealed more about their views on why the companies were formed.

All those interviewed recognised that university companies were no different from any other, in that ultimately they had to be profitable to survive and indeed 87 per cent claimed to use profitability as a measure of performance. Some however admitted that assessments of company performance were neither rigorous nor regular. There were differences of opinion about the appropriateness of profitability as a sole criterion and no specific return-on-investment yardsticks or payback times were offered.

Forty-seven per cent of universities, including many of those who claimed to use commercial measures of performance acknowledged that their companies had brought other, less tangible, benefits. Several contrasted their profitability expectations for companies on the one hand with those for other university-based research units and centres on the other, for which a much less commercial view had traditionally been taken.

The following were typical responses to the question - What criteria are used to measure your university companies' performance?

'... evidence of continuing connection with the College, evidence that the operation was still compatible with the original business plan presented to the College and evidence that the academic involved was maintaining his responsibilities to the College at an appropriate level. The Head of Department would be asked whether or not he was satisfied that this operation should be allowed to continue.'

'We don't do it professionally really. If each year we get more business and there is no, as it were, flak coming in from the University ... We paid 20 per cent dividend on paid-up capital and this year we are paying 25 per cent, so that's one criterion, the external investors. I think the College criterion is how much money are these people making for us ... if every year (the academics doing consultancy through a company) get a contribution to their salaries ...that's important.'

'... we would judge commercial success in terms of profitability and turnover in the normal way. We would also judge success on the valuation of our intellectual property that we have licensed into those companies. We hang on to it. We don't assign the intellectual property, we will license the patent rights and we form a view that ...sometimes even in our relatively unsuccessful company we can see that our intellectual property has grown much more valuable. It has been developed and become exposed ...and the big players begin to take an interest ... We also, in looking at our venture overall, see success not just in commercial terms but also in political and other income, less tangible terms.'

A much narrower view was taken by the production manager of a campus-based biotechnology company (which has since gone out of business) - 'It's products coming out. We grew last year, we had two people when we started ... As regards our yardsticks it is how many projects we bring to completion and ultimately how many we sell-on.'

Others, although recognising the legitimacy of non-financial criteria, were at pains to stress that these other measures were secondary to the need to be profitable.

'Ultimately, the company wants to make a profit, ... we are looking to investments to make profits such that they can pay us the dividend. That is the prime consideration. There is the temptation to judge success on employment and other criteria but that is a by-product of profitability.'

That comment may well have been a reaction against the great emphasis and importance attached to job creation in Northern Ireland, which at that time had the highest regional unemployment rate in the United Kingdom and where there is an understandable tendency to judge success in terms of new jobs created, but similar points were made by others - the following from a university in the more prosperous South East of England, for example,

'First of all (success would be judged) entirely as you would any other company, whether the time and effort is being paid for in terms of business. There is of course a separate assessment which incorporates things like - are we making lots of good new relationships through the company? But that is secondary.'

For various reasons many universities have shied away from making formal assessments of performance. Some said they felt it was too early to do so with embryonic companies but there was also a recognition of the particular sensitivities of academic entrepreneurs to being told how to run their companies. For example the Director of an innovation centre at one university expressed it as follows

'We leave the companies to get on with this (assessment of performance)... What we have is an open door policy and encourage people to come and talk to us when they have got problems but the problem there is they don't always realise that they have got problems or ... it might be too late because they are not experienced. On the other hand, because of the psyche of the people who are starting these businesses, there is a tremendous desire for independence and non-interference. They have left the University because they don't want to be interfered with ... they don't like bureaucracy and they don't fancy being told what to do. They are also very clever so they don't want to be told what to do by people who have got an IQ of about half theirs, how to run their business, despite the fact that you might have experience, so it's a very difficult balancing act between doing what you believe to be in peoples' best interests but yet actually taking them with you. ... It's a balance ... what we ought to do for any company that is under six months old is insist that we monitor their performance on a monthly basis as part of the package- we do not charge. We don't do that at the moment, we encourage them to come.'

Although 87 per cent of universities claimed to use commercial criteria to judge the success of their spin-out companies, only a few did so with great conviction and most recognised the importance of other less tangible benefits. Uncertainty about how universities should judge the success of their companies and the sometimes evasive answers given, suggest that the real reasons for company formation had not always been rigorously thought through. Some university administrations had different expectations from their academic entrepreneurs - this despite the fact that the latter had frequently been the prime-movers in persuading their universities to form companies to exploit their research.

Roberts (1991, Chapter 9) noted the difficulty of measuring success and noted that any measure of company performance is likely to be incompatible with the exact measure the entrepreneur has in his own mind. Nevertheless, he attempted with some success to develop measures which would correspond to how the public might judge the success of a young high-technology company. This was a weighted performance rating which took account of the average sales growth over the life of the company, the numbers of years the company had been in business and the profitability of the company.

Samsom and Gurdon (1990) have argued that cultural differences can exacerbate this type of situation.

When a shared vision is no longer present, conflicts arise and seemingly can only be resolved with the departure of one of the players or a change in position. These situations usually pitted a scientist against a businessman. ...These inherent cultural differences place further emphasis on the need to define the organisation's mission and objectives from the beginning and ensure commitment from all members to these goals.

Although Samsom and Gurdon were contrasting the different expectations of scientists and businessmen these same issues were apparent in the present study between researchers on the one hand and their universities on the other. One university illustrated this by citing a case in which a senior professor who had been the driving force in creating a biotechnology company was dismayed when the Board of Directors refused to approve expenditure on an expensive item of equipment for his academic research. This occurred at a time when the company had little or no sales revenue and the priority was to market its existing products. Despite having approved the company business plan and its objectives, it was clear that the professor's expectations were very different from those of the rest of the Board. He left the Board and the company in high dudgeon and as a result relationships between the academic department of which he was head and the company took many years to recover and then only after the professor's retirement from the university.

If there is a lesson to be learned here, it is that the question of how to measure performance cannot be separated from the objectives of the undertaking. The expectations of all parties to the venture should, if not identical, at least be overt and mutually compatible. Only then can performance yardsticks be agreed. The need for full disclosure of and agreement between all those involved about their expectations and motivations can hardly be over-estimated. This is not new, Drucker was writing about the need for clear objectives, strategy and tactics (OST) 40 years ago (Drucker, 1957). Profitability is not necessarily the main objective of a company. In 1965, Pat Haggerty who was then President of Texas Instruments Inc (TI) wrote

Texas Instruments exists to create, make and market useful products and services to satisfy the need of its customers throughout the world [and] the opportunity to make a profit is TI's incentive to create, make and market useful products and services (TI, 1965).

Nor is this an out of date concept. Handy (1994) asks - What is a business for? He concludes it is not about making profits, per se but to make a profit in order to continue to do things or make things better or more abundantly. This point is made in the present context because of a tendency for some in university administrations, and particularly those from finance departments, to dismiss the non-financial objectives of academics as being somehow less than valid or respectable. In urging that there should be an overt recognition and reconciliation of different objectives there should be parity of validity. Nevertheless, if objectives cannot be reconciled it is cheaper and less painful for one or more parties to walk away before a company is formed that afterwards.

Motivations of the Academic Founders

This question was designed to discover the motivations of the founders of the companies, as distinct from those of the university itself. However, since it was asked in most cases not of the founders themselves, there is bound to be a degree of speculation, albeit by people close to the companies and their founders. There are recognisable patterns in the replies across institutions and the motivations attributed to the researchers which coincide with those reported in other studies (Roberts, 1991; Olofsson et al., 1987).

Not surprisingly there was a variety of answers and many gave more than one reason but there were three predominant themes:

- The desire of researchers to see their work exploited commercially for its own sake - 81 per cent of respondents;
- To get funding for academic research - 25 per cent of respondents;
- To make money - 63 per cent of total but only 19 per cent gave this as their main reason.

The following are some examples of the desire of researchers to see their work exploited commercially.

'In the first case, desire to bring modern technology into his teaching in a more effective way, and using a vehicle which had good development possibilities in other areas and a desire to see his discipline more actively involved in Irish life in a very genuine sort of way.'

'... a research opportunity which wasn't being availed of properly which he took over and developed into a business.'

'... output from a research programme which normally in the University context would have died but the individual could see that it had possibilities outside the University context and carried it there.'

'In the case of (company name) they wanted to get the product into the market. In (another case) I think it was the profile of the academic department that was important, the profile of their work rather than the motivation to make a lot of money.'

Some saw company formation as a way to obtain funding for ongoing and future research.

'Professor ... was keen to get his product out to market. ...but he would just as quick want to see (company named) pay for R&D in his Department now, rather than simply benefit personally.'

'... there's one here which was motivated entirely by the desire to earn research funds by providing expert services to companies.'

'... firstly to fund research and ...other things at the same time. Secondly, ... to get some consultancy out of it , although they were never put like that and the third thing is that they can also bring and involve students, especially graduate students in that research, which is (of) considerable benefit.'

'A variety (of motivations) ... financial is clearly one. I think people genuinely want to see science used or they want to see science and technology exploited in the broadest sense and see the opportunities when they arise. I think there is a higher awareness among academics of those possibilities than sometimes (they are) given credit for.'

'Primarily to see their work in use. ... their end-points are not just the publications ... it is actually making and selling and getting, I suppose, acclamation of the public in the market-place.'

'... in my experience the large majority of academic staff give that (getting rich) a fairly low priority. I'm surprised at the numbers who if they get rich through licences or whatever will in fact put a lot of the money back into research funds in the University, to further their own work. I'm always pleasantly surprised about that.'

Although it was believed that most researchers thought that making money was a relatively low priority, only 18 per cent ranked it as the main motivation, it was cited as one of the driving factors by 63 per cent of those interviewed.

'They want to develop technology and have a ... comfortable lifestyle and they want to have the freedom to do what they want to do and want to be able to pay the bills at the end of the week and have a smile on their face for most of the week.'

'Because they saw a commercial opportunity to make some money, simple ... and they reckoned that was the way to do it.'

'The desire to actually get something financially out of it.. That's it.'

'... There is a feeling that ... other people are going to make money out of my work, why shouldn't I try to make some of this money myself?'

One research park Director felt that not wanting to be rich, or at any rate not admitting to wanting to be rich, was a positive disadvantage when dealing with venture capitalists who had no such inhibitions.

'... people are very reluctant to come out and say they want to be very rich. Of course that's one of the fundamental problems when they talk to venture capitalists because the venture capitalist has no hesitation - he wants to get very rich and he's putting up most of the money - and this is one of the difficulties. So whether it's coyness or whether it's genuine that these people simply would like to see this thing go through and be successful and give them the sort of independence that comes from that success - but really don't have any compelling desire that the thing should be a really large operation - is very hard to determine. I really don't know what motivates them.'

The interview results in the present research are in broad agreement with previous studies. The findings of Samsom and Gurdon's (1990) Canadian study were that the main motivations of 22 scientist/entrepreneurs were:

- Advancement of science and its
  application   86 per cent of respondents;
- Personal opportunity to build a business,
  to become an entrepreneur   82 per cent of respondents;
- Opportunity to build equity,
  make money   55 per cent of respondents.

Direct comparison with Lumme's study of firms in Finland and Cambridge is difficult because of the ranking system used by Lumme but the most important reasons given by researchers for forming companies were the 'Desire to develop one's own ideas', closely followed by

'Identified customer needs or identified defects of existing products', 'Top-class technological know-how', 'Personal motivation to succeed' and 'Good marketing ideas for the product'. All of these scored between 2.2 and 2.5 out of a maximum of 3 on Lumme's ranking scale. 'Attempt to raise one's income level' scored 1.4 amongst the Cambridge entrepreneurs and did not feature at all in the Finnish sample Lumme et al. (1992).

Although these studies revealed a great diversity of motivations the main drive in all cases has been the desire of researchers to see their work exploited. In all of them the desire to make money figured to some extent for its own sake, 63 per cent in the present study (although less than one fifth gave this as the main reason), 55 per cent in the Samsom and Gurdon study while in Lumme's sample it was ranked fifteenth out of 27 factors scored by the researchers. According to Gorman et al. (1988) the main impetus for entrepreneurs starting NTBFs were personal accomplishments and peer recognition through publication of new findings. Earning money was not the main reason. The only significant additional feature in the present study is that a quarter of the university respondents believed that one of the main motivations of their researchers was to obtain research funding. This may be a reflection of the financial difficulties which the universities were facing in the early 1990s.

Significantly these are all what Smilor et al. (1990) referred to as 'pull factors', there was little evidence in this sample of 'push factors', as reported by Weatherston (1994), resulting from frustration with their universities or a desire to abandon academic life. Weatherson's study of 26 academic entrepreneurs from 12 universities showed that although they maintained close links with their institution, they spent only a small proportion of time working in the company and most did not want to enter business life full-time. This it was concluded, would be sure to have an adverse effect on company growth. This latter conclusion is confirmed in the present research in which the importance of full-time managers emerged as of great importance.

In summary, as was found by most previous researchers, pull-factors predominated and although individual instances of frustration surfaced during the interviews in the current research there was no evidence in any of the universities visited that academics had been driven to leave academia to form companies. On the other hand most of the interviews were with universities which had taken a stake in their companies, implying at least a financial interest in their success. There might well be disenchanted academics elsewhere.

This apparent enthusiasm of significant numbers of academics to see their research exploited commercially and to take an active part in the process is encouraging. The alternative would be bleak. Etzkowitz (1983) has pointed to the danger that if research scientists choose to restrict communication with industry they can do so but it is unlikely that restrictions will come from industry.

> ... therein lies the paradox of entrepreneurial science. If basic research is put aside because of commercial opportunities, it will be the result of scientists' choices. Rarely have a social group had such freedom to establish their own course, and to set the terms on which they will receive the resources to pursue it, as do contemporary academic scientists.

Many, in industry as well as government, would view the prospect of economic growth being regulated by academics in this way as a recipe for a disaster which could and should be averted. An understanding of the motivations of would-be academic entrepreneurs and the factors which foster or inhibit commercial activity is a prerequisite to the design of policies to encourage them.

Conflicts of Interest and Difficulties of Transition

A number of researchers have explored the tensions, both real and imagined, experienced by researchers who become entrepreneurs and the conflicts of interest between their academic and corporate roles (Etzkowitz, ibid.; Samsom and Gurdon, ibid.; Preston, ibid.; Bullock, ibid., Roberts, ibid.; Louis et al., 1989). These have mainly concentrated on difficulties of transition, to attitudes and perceptions and on practical considerations such as the conflict between the desire to publish on the one hand and to protect intellectual property for commercial ends on the other.

There is however another type of conflict which occurs as a result of the structures and organisational arrangements, especially those relating to the management and direction, of university companies. This affects university staff who are also directors of companies and particularly university administrative or industrial liaison staff, part of whose university role is to maximise research funding from industry. If 'industry' now includes spin-out companies in which the university has a stake, the dilemma is obvious.

Twenty-nine percent of university administrators admitted to having experienced such conflicts of interest. Some responses to the question of whether conflicts of interest had been experienced illustrate the experience of practitioners.

'Yes I am aware of conflicts. To be specific I am a director of two of the companies, in one case by decree of the University authorities and in the other by invitation of the promoter... The conflict of interest is of course one which would only ultimately be resolved by resigning from the company if one thought that its actions were totally hostile or inimical to the general College's welfare. ... I haven't personally had any conflict of interest problems that I couldn't resolve quite quickly.'

'There is a potential conflict of interest, yes, which I have to be careful about between my University role and my role in the companies. One or two companies than I am on the boards of rely on me from time to time ... that I am there in a personal capacity and not as mandated, because I can't be, the Companies' Act doesn't allow it. There is a potential conflict of interest which I would have to resolve as it occurred.'

Another difficulty which can arise, but one which several universities have recognised and taken pains to deal with, is that in which a company competes for research or contracts with the university.

'Yes when looking at it from a University perspective whether I should take certain regulations or certain activities for the reason ... that perhaps it (a piece of work) should be better done through the University. On the other hand if you are actually sitting on the Board of a company you are thinking of the Company's perspective and that is where the difference would lie. I'm not saying there have been tensions at Board meetings. I'd say the tension is more outside ... in terms of what decisions or recommendations you should make.'

'I don't find that (conflict of interest) a problem. I have to say that there are occasions when I am negotiating for both sides at once. ... I don't think that is conflict of interest, it is a potential conflict of interest situation because on the one side I could be saying in the Company I want my pound of flesh, my ten per cent and on the other side I could be saying that I can't afford it. There are those situations but they always get resolved. It would be much more difficult actually if we had two people running it separately, because then there would be different interests.'

'Yes, I am very aware of the conflict of interest question and I take considerable pains to avoid it and I also take considerable care to ensure that my academic colleagues also avoid conflicts of interest. This has been discussed (and) some guidelines have been laid down (by the University) ... because if you are in two companies - as director of two companies - then the conflict of interest is covered by the Companies Acts. If you are in a company and also in a university as a senior employee, the Companies Act doesn't apply but the morality does.'

Ideally, university-company structures should try to avoid placing staff in such difficult roles but in practice there are often good reasons why university staff should serve on company boards. To prevent them doing so would in many cases hurt both the university and the company but the least that can be done is to recognise the potential for divided loyalties and bring any problems which arise into the open.

The responses suggest that most of those involved recognised the conflict of interest problem but all claimed to be able to live with it. There is no way of knowing whether this should be taken at its face value or if it is simply a rationalisation by well-meaning people to avoid having to resolve a near intractable problem. There does seem to be an inevitability that sooner or later tensions will arise but provided they are recognised they can be dealt with, as John Preston of MIT observed

> ... the only way to totally avoid conflict is to stop both university relations with industry and the entrepreneurial activity of university researchers (Preston, 1992 a).

Turning now to problems experienced in the transition from academic to entrepreneur, although many acknowledged that there had been difficulties, most seemed to have been able to cope fairly philosophically. Several of those interviewed referred to a 'me-too' factor in which academics who see their colleagues taking the step of forming companies want to emulate them. This is something more than a desire for financial reward as it has been noted in situations where to date there has been no financial return to those involved. It may simply be that once their colleagues have taken the step and demonstrated that this is an acceptable and respectable activity for an academic, others will follow. Whatever the explanation this is an important phenomenon. It is not quite a 'snowball' effect but certainly a number of universities which found it difficult to attract academic entrepreneurs when they began their company formation activities now find many more coming forward with ideas for exploitation. This effect was remarked upon mostly by those universities which attempt systematically to commercialise their research.

Influences of Market Pull and Technology Push

Interviewees were asked for their views on whether each of their university's companies was formed primarily as a result of market demand or whether they were driven by the technology i.e. market-pull or technology-push. This question is of course a deliberate over-simplification given the frequently complex process of innovation. Nevertheless it produced some unexpected responses. One-fifth of universities claimed that they had equal numbers of companies formed mainly as a result of 'market-pull' as of 'technology-push' but the remaining 80 per cent believed their companies to be overwhelmingly technology driven. None claimed to have had totally market-driven start-ups.

The question is not perhaps as straightforward as it seems. One problem is that there is no common understanding of market-pull. Market-pull was interpreted in some cases as something for which someone or some group of potential customers had specifically asked and several companies can be identified as having been set up specifically to satisfy that kind of well defined requirement. Others however failed to distinguish between market needs, as believed to exist by the entrepreneurs, and wants which had actually been expressed by potential customers. In the opinions of those interviewed some academic entrepreneurs seem, rather naïvely, to have assumed that demand would naturally follow the commercialisation of their ideas - a kind of 'better mouse-trap' syndrome.

Sometimes of course the entrepreneurs got it right, demand was created and a number of successful companies have been formed as a result of this sequence in which new developments have exposed latent needs and created new markets. With hindsight it is tempting to call this market-pull and it is arguable that at some stage technology-push shades into or is converted to market-pull.

If all of this were simply to demonstrate that this was a difficult question, it would be rather pointless. There are however some serious issues. No company can grow without a market for its products and if university-based entrepreneurs are good (or lucky) at picking winners they have the opportunity to create demand for new products or meet existing demands in novel ways. Being first with innovative products, particularly if associated intellectual property can be protected, offers great opportunities and this fact alone is probably sufficient motivation for universities to speculate on their technology. Unfortunately however not everyone backs winners and the chances of failure are almost bound to be higher in such ventures that in those which set out to supply an articulated and quantifiable demand from identified customers or users.

Despite anecdotal evidence of the superiority of the market-driven approach - most frequently advocated by marketers - the case is not proven. There are a great many examples of technology-driven firms which ultimately become extremely successful and 80 per cent of the spin-outs in this survey are believed to be the result of attempts to commercialise technology. It may in fact be one of the distinguishing characteristics of university spin-out companies that they more frequently begin as attempts to exploit research that from well researched market needs. It may also be a strength of this unique environment for start-ups that such ventures are more likely to be encouraged and nurtured than they would be in the sometimes harsher commercial world. One might also speculate that, because as already noted other less tangible benefits are recognised and valued by the university, there may be a greater tolerance of slow growth and a longer time to reach profitability than would be the case if only return-on-investment criteria were to be applied.

The reality may well be that the extreme conditions of market-pull and technology-push are caricatures of what happens in the real world and the point at which technology-push shades into market-pull is not always recognised. The terms market-pull and technology-push are ill-defined and frequently misunderstood, they have almost become clichés and are, in fact, of little help in this context. They oversimplify the innovation process and perpetuate the myth, beloved of marketers, that everything must be market driven.

With rare exceptions most university spin-outs have occurred as a result of Rothwell's third or fourth generation processes, described earlier. They have either reached, or are attempting to reach, the situation where a market has been created but this has by no means been a linear or a steady progression. The route to successful innovation is often only clear in retrospect. Although eventually a market for a company's products is essential for commercial success, neither universities nor academic entrepreneurs should be deterred from exploiting new technologies.

The Role of Entrepreneurs and Product Champions

This question was asked to test the widely held view that, to be successful, new companies require a dedicated and determined individual who is totally committed to making the enterprise succeed. This person is sometimes described as a product champion, sometimes as an entrepreneur. Ideally these people will have tenacity and staying power - or stubbornness - to overcome obstacles which would defeat others and as someone put it, be prepared to climb walls with their fingernails to make their ideas work.

The distinction between entrepreneur and product champion was deliberately left undefined and, not surprisingly, respondents had differing understandings about these roles. For a discussion of the role of the entrepreneur in high-technology companies see Oakey et al. (1988). The distinction between the roles of entrepreneurs, inventors and champions has perhaps been best described by Stankiewicz (1986), already quoted in Chapter 7.

Seventy-nine per cent of interviewees were able to identify some dominant individual who could be described as an entrepreneur who had played a leading role in the success of each company. Most also recognised the role of product champion as being different and although sometimes one person performed both sets of functions, in most cases the product champion was someone other than the entrepreneur. Fifty-seven per cent felt that a champion was a sine qua non for company success.

An example '...we will only accept it (a proposition to form a company) if someone from the University will champion it, so by definition there has to be a product champion..' Others had to think more carefully about it 'I think in every case there is an individual, yes ... if there isn't a proper project champion, they just won't work I don't think. It's difficult to make these things happen.'

In a few cases the roles of entrepreneur and product champion were regarded as interchangeable but others made a sharp distinction. The director of a university technology park/innovation centre was quite clear about the difference.

'No, there are no obvious entrepreneurs in this building in the sense of the London barrow-boy type ... There are people who are more commercial than other people. There are very few people however with a sharp, commercial edge. I would say that there are about two or three in the building who are truly commercial. They are all totally committed to what they are doing. They are not product champions in that they are boring about it and they have nothing else to talk about or that you are fighting the world with your product but in total commitment and the number of hours you put in to actually working through and getting your product to market, they are working on average 60, 70, 80 hours a week, that sort of order and that is because they love their product.'

A number of universities were able to identify companies which had product champions, mostly drawn from academics who understood the technology involved, but which lacked entrepreneurs. In these cases the universities had made attempts to fulfil the entrepreneurial role by other means, for example through a holding company or a partnership with existing companies.

Several respondents observed that if a project hadn't got an obvious entrepreneur it was necessary to recruit one. In one group of 12 spin-out companies, ten had identifiable product champions before their formation and one of these also proved to be a very effective entrepreneur. In five companies the initial entrepreneurial moves were made jointly by the university holding company in collaboration with existing, larger companies - a form of corporate venturing. Nevertheless, in nine of the companies it was necessary to recruit key people with entrepreneurial ability at an early stage. One company which failed to do this soon enough found itself in the doldrums.

Perhaps the single greatest factor affecting the success of spin-out firms has been the presence of key individuals. Often the most influential person is the founder who may be the kind of charismatic individual who can enthuse and galvanise others. Sometimes this person is referred to as an entrepreneur, and although there was no consensus amongst practitioners about defining exactly what this means, most claim to recognise when the role is not being fulfilled. In which case it is usually necessary to recruit someone to meet that need. Frequently too, a product champion can be identified as crucial to success and in university-based firms this person has often emerged from the ranks of the researchers.

Finally, there is a need in all cases for a manager and everything points to the need for this to be a full-time post filled by someone who is prepared to make a personal commitment to the company's success. Although many of the companies have started out with academics in the role of part-time managers this has seldom been a satisfactory situation as the firms grow. The occasional exception to this was in the case of 'soft' companies, such as part-time consultancies, which are little more than firms in name only and which required little management expertise, relying solely on the technical abilities of the academic staff involved. No case was identified in which a founding academic had successfully fulfilled the role of Managing Director or General Manager on a part-time basis beyond the embryonic stages and the experience of most practitioners was that it was best to use the expertise of academics as technical directors or consultants rather than to employ them in executive roles[2].

In a study of 22 founders of high-technology-based firms in Boston, Gorman et al. (ibid.) found that although most felt that their company start-up had not been particularly risky, their concerns had been more to do with their possible loss of personal reputation or prestige amongst their peers than with failure of the company. About 70 per cent of these academic entrepreneurs remained employed at least to some extent by their previous university after having formed companies. Many admitted to having had difficulties related to managerial aspects of running their companies.

It is possible that the roles of entrepreneur, product champion and manager could all be performed by a single individual but in practice it is unusual for this situation to persist satisfactorily beyond the embryonic growth stage. In an American study of technology-based firms Miner et al. (1989) concluded that entrepreneurs were less growth-oriented but more task motivated than managers. Furthermore those entrepreneurs searching for financing from venture capitalists had significantly higher task motivation than those who were not seeking funds from venture capitalists. Numerous instances were described during the survey in which a company's needs had outgrown the ability of its founder or founders to manage it effectively.

It seems to be the case that while product champions are likely to 'emerge', mainly from the ranks of a university's researchers, entrepreneurs are much more likely to have to be recruited. A more precise description of the latter role and of the characteristics required to fulfil it would be desirable to help with selection and recruitment. Although there were differences in definition and in emphasis about their respective roles, all of the universities interviewed believed that the roles of entrepreneurs, product champions and managers were essential for success.

Original Expectations in Hindsight

The interviewees were asked whether, in hindsight, the original expectations for the companies had been met. Experiences were almost equally divided between those companies which were felt to have done well and those which fell short of original expectations. Bearing in mind the earlier observation that there was some ambiguity in many universities of how success should be measured and that many of those interviewed did not have ready access to the financial accounts of their universities' companies, this question was in most cases answered subjectively. The results are nevertheless valuable, since it was the perceptions of the universities which were being sought. A more objective approach to measuring company performance is taken in later chapters, see Chapters 12-15.

Three universities in particular claimed to have exceeded their expectations and several people stressed the difference between their personal expectations, which was the question as put to them, and the different expectations of others.

'Well, my expectations as distinct from other peoples' say the Board's expectations... I didn't expect to see all that many companies formed or indeed that it would be all that successful. My background told me that it was a very tough wicket and I didn't expect...certainly I never envisaged ten or eleven companies after five years ...'

'... we exceeded our expectations because we planned on subsidising the company with two per cent for six years and found we could cut it back to one per cent subsidy in three years. In that sense we exceeded our expectations and the technology-transfer income is now beginning to appear.'

It is no surprise that the industrially experienced and worldly-wise had less extravagant expectations than the academics involved but there was some frustration amongst the former group that their academic colleagues were slow to learn the commercial lessons.

'I think I am disappointed in the speed with which some of our promoters and operators have learned the obvious lessons from what is happening around them but I don't think they are significantly behind where I expected them to be. They are about where an objective outsider would expect them to be, bearing in mind the kinds of people they are and the kind of things they are doing.'

'Expectations were for them to grow easily, to transfer to independence relatively easily and to bring us a big return. All those expectations were not met ... tremendous problems!'

There was, too, some genuine disappointment at failure of some companies to achieve what was believed to be their reasonable potential.

'I think a lot of them have under-achieved really. I don't know what the expectations of the founder members were but looking at it as an outsider I think a lot of them have under-achieved. There have been very few companies that have grown to anything like a substantial size. I think a lot of them would have expected to grow bigger.'

'I think there has been a sense of disappointment. Not a disaster but I think that we were a little over-optimistic. ... we well understood that if we had one big winner out of ten goes then we would be content. I think that we started by backing ventures that were too big and too expensive. We put a rather small number of large eggs in one basket so that difficulties on that front were quite painful. That wasn't policy it was just circumstances....'

The Director of a Science Park felt it was really too early to judge and companies formed during the recession were having a particularly difficult time just to survive.

'It's too early days because we have only been going four years. Most peoples' horizons are probably seven to ten years. They want the company (to be) in the golden uplands, at the stage where they don't have to work 80 hours a week but they can still develop the technology and can earn a reasonable reward without killing themselves. That seems to be what they want. What has happened recently is that the recession has held them back in that because we are all having to try and survive so it is a bit difficult to try and answer your question specifically but I see no evidence of anyone who wants to commit suicide. In other words although it is very tough in the real world, they are coping.'

With some notable exceptions, more disappointment than satisfaction was expressed. It has to be remembered of course that many of the universities were relatively inexperienced and were speaking of expectations which they and others had at a time when they were embarking on something outside their previous experience. It is likely that their initial expectations were unrealistically high. Few universities had more than five years experience to draw upon and no doubt they would be less sanguine or naïve about future prospects.

There is a real danger that, in their enthusiasm for new company formation, universities can raise unrealistic expectations amongst their staff or in the community generally about what can be achieved. This is perhaps less likely to occur now than a few years ago when a university which was forming companies was more newsworthy and when some extravagant claims were made about the potential for employment creation and profits.

Significant Factors which have Shaped the Companies

What Factors were Most Significant in Shaping the Companies?

The practitioners were asked what three factors, in retrospect, had greatest influence on their universities' companies. Although no clear pattern emerged, many respondents referred to the influence of the founders of the companies and other key individuals. Some examples,

'They have been enormously dominated by the one individual who pushed the thing forward in the first instance and who has hammered out a shape which only the leading company has managed to escape from. Only the largest one is shedding the image which the leading man created and that is because he took a conscious decision to do that. All of the others are entirely creatures of the personality of the guy who went in.'

'The individual champion of the technology, that's usually the technical man. The support we've had from a particular lawyer, a friend of the University not one that we have to pay for, in structuring the shape of the company. I think also the Vice-Chancellor has been very supportive.'

'... the fact that the person happened to be in a position in the University ...to utilise the University resources to allow that company to get up and running.'

'It's the competence of the academic entrepreneur and his persistence in overcoming all sort of problems which is the key factor in getting these companies off the ground. ... Overall the environment ... is generally supportive, as supportive as it can be in a UK situation of providing help. Networking around here is very important - people networking ... is ... a very important way of helping these people survive and feel they are not in isolation. Where we have problems with academic type business it's because the academics are very good at talking and very bad at listening and have the wish that the world was not the world that they are in - an inability to relate to the world.'

'I think the selection of the chief executive in my opinion has been the most critical factor in terms of company success or otherwise. Secondly, it's the ability of the initial sponsors, investors to reliably come forward with further rounds of funding when required, i.e. staying power. And I think thirdly it's getting the role of the academic entrepreneur sorted out early.'

'... the ability of the company to carve out new markets for the particular skill or product which they started off with and to almost haul themselves up concrete walls where no toe-holds seemed to be there, really gone for things which perhaps more sober people would not have attempted.'

These views confirm the critical importance indicated in the literature of the roles of entrepreneurs, product champions and managers. The universities' attitudes and commitment also figured widely.

'I think the University's position with a very hands-off role has been crucial ... I guess things like the Science Park have been very influential because it's provided both a facility and its also raised the level of consciousness of the possibilities of this sort of activity and I think that has been very important. And I think there was a financial resource as well .. and not just money but networks.'

'The University itself has this long-standing tradition of involvement with industry ... That's been influential. I think the ability of the Science Park ... has actually shown the possibility of being located.. very nearby and still treated as part of the campus community.'

Other respondents referred to the quality of the academic resource, the importance of becoming commercial, achieving credibility and becoming financially self-supporting.

'The reason the majority of them have come here is because of the academic connections, that's the plus. The fact that we have the space is the second reason.'

'I think the high quality of the intellectual resource. The willingness of the University to accept the principle of having University companies. And ... on the negative side, the fear of risk because of having such small financial reserves.'

'At the beginning they were shaped by our needs to get something off the ground for credibility. Later on they were shaped by our need to be totally commercial and use partners and external influences to enhance the companies. Those would be two key factors in shaping the companies. ... the fact that we had limited financial resources meant that it constrained the shape of the companies.'

'Well certainly in the case of one of them it was the seed money, where it came from and the deal that was struck with the industrial partner in terms of support services and overseas marketing has obviously shaped that company.'

'Hard work. ... I'm confronted with more opportunities to do things than we can handle. Probably, even more important than that, the attitude of the senior management. The third one, to be able to call on people on the Board of the company ... these sort of guys have got a different attitude towards what we are doing than the academics do and this is healthy.'

Rating and Ranking of the Importance of Various Factors

Interviewees were then asked to rate the importance of the following list of factors to the success of the Companies, using a scale of 1 to 5; from 1 = not important at all, to 5 = very important:

- Marketing;
- Technology;
- Patenting or otherwise protecting IPR;
- Public relations, within university?, externally? ('hype');
- Entrepreneur(s);
- Product champion(s);
- External partners;
- Share ownership by inventors/researchers;
- Inventors/ researchers role, what it should be;

- Relationship with university departments;
- Relationship with wider university system and administration;
- Sources of funds;
- Full-time managers;
- Composition of Board of Directors;
- Making profits;
- Return on investment;
- Rewarding inventors/researchers;
- Anything else?

Respondents were asked to give their views on the importance of these factors. They were deliberately asked to score the factors based on their initial reactions without being given time for detailed reflection. Figure 8.1 and Tables 8.1 and 8.2 display the results. This question was also put, during separate interviews, to 24 managers/entrepreneurs of university spin-out firms as well as to the university staff. The results are combined here, to give a total response of 44. A few significant differences in the responses are referred to in the text.

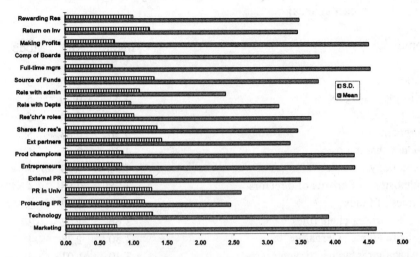

Figure 8.1 Relative importance of factors - means and standard deviations (n=44)

Table 8.1 is sorted by decreasing values of mean score to highlight the factors scored most highly while Table 8.2 is the same data ordered by increasing value of standard deviation. This gives an indication of the areas of greatest agreement. There was wide agreement about the importance of some of these factors but considerable differences about others. Comparison of the means and standard deviations reveals that attention to marketing, being profitable and the importance of full-time managers scored highly. None of the respondents scored any of these factors at less than three. The need for entrepreneurs and product champions was also widely accepted confirming the opinions expressed during the interviews.

Table 8.1   Scores of 44 respondents (sorted by descending order of mean scores)

Topic	Mean	SD	Mode
Marketing	4.61	0.75	5
Full-time managers	4.53	0.70	5
Making profits	4.51	0.74	5
Entrepreneurs	4.30	0.83	5
Product champions	4.30	0.85	5
Technology	3.91	1.29	5
Composition of boards of directors	3.78	0.88	4
Sources of funds	3.77	1.32	5
Inventors/researchers roles	3.66	1.02	3
Public relations externally	3.50	1.28	4
Rewarding researchers/inventors	3.49	1.01	4
Return on investment	3.47	1.26	4
Share ownership by inventors, researchers	3.46	1.36	4
External partners	3.35	1.43	5
Relationships with university departments	3.18	0.97	4
Public relations more widely within the university	2.61	1.28	3
Patenting or otherwise protecting IPR	2.45	1.17	3
Relationships with wider university system and admin	2.39	1.10	3

There was less consensus about the other factors and greatest differences of opinion occurred about the importance of sources of funding, having external partners and of share ownership by researchers. That is not to say that the latter considerations are unimportant, some universities scored them highly but there was less consensus. Lowest on the list were the perceived importance of maintaining good relations with the administration and wider university system, the importance of patenting or otherwise protecting intellectual property and concern for PR for the companies within the university.

There were however a few significant differences between the responses of the university staff and the entrepreneurs interviewed. These are best observed by comparison of the modal scores of each group for the topics on which they differed, see Figure 8.2.

Table 8.2 Scores of 44 respondents (by ascending order of standard deviations)

Topic	Mean	SD	Mode
Full-time managers	4.53	0.70	5
Making profits	4.51	0.74	5
Marketing	4.61	0.75	5
Entrepreneurs	4.30	0.83	5
Product champions	4.30	0.85	5
Composition of boards of directors	3.78	0.88	4
Relationships with university departments	3.18	0.97	4
Rewarding researchers/inventors	3.49	1.01	4
Inventors /researchers roles	3.66	1.02	3
Relationships with wider university system and admin	2.39	1.10	3
Patenting or otherwise protecting IPR	2.45	1.17	3
Return on investment	3.47	1.26	4
Public Relations more widely within the university	2.61	1.28	3
Public relations externally	3.50	1.28	4
Technology	3.91	1.29	5
Sources of funds	3.77	1.32	5
Share ownership by inventors, researchers	3.46	1.36	4
External partners	3.35	1.43	5

The importance of maintaining good PR for the companies within the university generally was more highly rated by the university administrative staff than by the academic entrepreneurs. By contrast the latter, perhaps because of their closer involvement with individual departments, were more concerned about maintaining good relationships with university departments than were the administrators. During the interviews numerous examples were quoted, both of difficulties encountered and of problems avoided, to illustrate the need to maintain good relationships with both constituencies More surprisingly the entrepreneurs rated both return on investment and the need for external partners more highly than did the administrative staff.

The relatively low score by the university administrative staff on the importance of return on investment criteria may reveal a more realistic appreciation of the time taken for embryonic firms to become profitable.

The administrative staff interviewed, were after all mostly industrial liaison people or others with business experience. There may also have been a recognition by those closest to university companies in general of the importance and validity of non-financial criteria, referred to earlier.

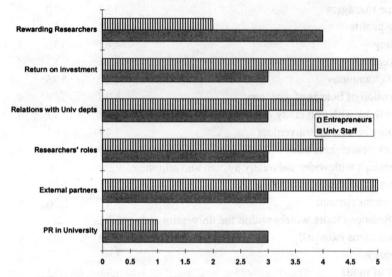

Figure 8.2 Modal scores of university staff (n=20) and entrepreneurs (n=24)

The differences in rating of the importance of external partners is more difficult to explain. The need for external partners may have been interpreted differently by the two groups. The question was put to the university staff at the conclusion of a discursive interview where these matters had been discussed in detail and during which their opinions had been sought. By contrast the questions were asked of the entrepreneurs after a more structured and factually based interview. As a consequence, the term 'partners' may have been interpreted differently by the two groups - a weakness in the questionnaire.

It is particularly interesting that the researchers/entrepreneurs were less concerned about the importance of rewarding researchers and inventors, i.e. themselves, than were the university staff who seemed to be almost over-sensitised to the need to be seen to reward academic staff. However, although not overly concerned about tangible rewards, the researchers did see their role as being an important ingredient for success.

This result tends to confirm the conclusions which have been drawn about the motivations of academic entrepreneurs, for most of whom tangible rewards were less important than seeing the results of their work exploited for its own sake.

Summary

A series of structured interviews with staff at universities which had collectively formed more than 200 companies revealed a great variety of types of company and forms of organisation to deal with them.

A number of critical issues were of concern to all, including:

- How decisions to form companies are taken;
- How company success or performance is judged;
- The relative importance of market-pull and technology-push;
- The importance of entrepreneurs and product champions;
- Motivations of universities and of academic entrepreneurs;
- Conflicts of interest for university staff;
- The effect of external factors on company formation and operation;
- The extent to which, in hindsight, original expectations had been met;
- The relative importance of various factors to the success of companies.

The growth of university companies has been the result of a number of pressures, not least the emergence of the enterprise culture of the 1980s. The most frequently expressed reasons given by universities for forming new companies were, financial pressures from government, 92 per cent; scarcity of research funding, 75 per cent; the proximity of a Science Park, 58 per cent; activities of other universities, 42 per cent; regional employment incentives, 33 per cent and availability of government grants, 25 per cent.

Differences in legislation regarding ownership of intellectual property has meant that in the UK the responsibility for exploitation rests with the universities whereas in USA and most of the rest of Europe the impetus is more likely to come from the academics who own the IPR in their research. Despite this there have been many instances of UK researchers taking the initiative to persuade their universities to form companies to exploit their work. Indeed some universities seem almost to have been 'bounced' into forming companies without due regard to alternative means of exploitation such as licensing. In some cases the universities have realised too late that the expectations of their academic entrepreneurs were at odds with those of the university. Many universities were vague about criteria to assess the performance or success of their companies, and a number admitted that the reasons for company formation had not been adequately thought through beforehand.

The motivations of the researchers as reported in this study are in broad agreement with most other research. Pull, as opposed to push factors predominate although the caveat must be made that most of those interviewed were from universities which had adopted policies of assisting or at least not discouraging academic entrepreneurship.

Unsurprisingly, the motivation for academic researchers to form companies were many and varied but there was nevertheless a recognisable pattern. Amongst a wide range of motivations were the apparent desire of researchers to see their work exploited commercially for its own sake, 81 per cent; to get funding for academic research, 25 per cent; to make money, 63 per cent in total although only 19 per cent gave this as their main reason. It has now become respectable for academics to become entrepreneurs and the emergence of a 'me-too' effect has been observed in a number of universities included in the survey.

Conflicts of interest were not perceived as being serious and those who admitted to experiencing tensions believed they could cope with them. Whether this is in fact the case is a matter for conjecture and one could imagine situations in which it would be difficult for university staff to remain totally objective. Nevertheless the probability of serious problems arising appears to be low and there are good reasons for not excluding university personnel from company boards and policy making.

More recent tensions have arisen from the means by which university research is now being funded in the UK being at odds with exhortations from government for universities to engage in industrially relevant research.

Although the terms entrepreneur and product champion were interpreted differently by different respondents, all agreed on the importance of key individuals and of the need for a balance of skills both at executive level and on the Boards of Directors. There was also broad agreement of the necessity to have full-time managers in university companies.

Relatively little attention appears to have been paid to the effect of external environmental factors of the companies although in hindsight it is clear that these were significant influences on the companies' success. The main factors in the external environment believed to have greatest influence on the decision to form companies were regional factors, university policies, government encouragement and assistance and lack of government funding for universities.

The point at which technology-push shades into market-pull was not always recognised until after the event. The complexity of the innovation process was generally recognised and simple terms such as technology-push and market-pull were imprecise and unhelpful other than as very broad descriptors. The vast majority of universities had examples to quote which demonstrate that, although to survive a company must have a market for its products, universities should not be deterred from speculating on their research. This is especially true when they can protect their IPR.

Some universities expressed more disappointment than satisfaction with the actual progress achieved when compared to their original expectations although many now admit that original hopes were unrealistic. There was disappointment on the part of some university administrators at how slow their academic colleagues had been to learn the commercial lessons.

Invited to rate on a scale of one to five the importance of a range of factors to their companies a total of 20 university staff and 24 academic entrepreneurs the highest scores were for marketing, being profitable, having full-time managers, and the need for entrepreneurs and product champions. Significant differences between the administrators and the academics were confined to the importance of maintaining good PR within the university which was scored more highly by the administrative staff; good PR with academic departments was regarded as more important by the academic staff. Two factors - a return on investment and having external partners - were also rated more highly by the academics.

Notes

1. This is a reference to grants from IDB and LEDU in Northern Ireland. IDB is the Industrial Development Board for Northern Ireland, LEDU is the Local Enterprise Development Unit - both government economic development

agencies. These agencies operate as part of the Department of Economic Development for Northern Ireland and undertake some of the functions carried out by DTI in Great Britain.

2. There have been a few examples of founder academics giving up their university careers to become full-time businessmen. For example the founders of London City University's successful gas sensor company City Technology Holdings which was recently capitalised on the stock market (The Times, 1996).

9 Characterisation of University Companies

Introduction

A striking feature of the interviews was the diversity of types of company, ranging, in Bullock's terms, from soft to hard and of the extent of involvement by the universities both as institutions and through their academic staff members. There were differences both between universities and even between different companies at particular institutions. A variety of different procedures and structures had been put in place to deal with company formation and support and to manage the interactions between the companies and the universities. Despite this there was a great deal of common ground and broad agreement amongst the practitioners about the importance of a surprising number of factors, some helpful to and others which inhibit the success of university-based companies. Although the same issues and considerations were highlighted time after time, the various universities had adopted widely differing ways of addressing them - some more successfully than others.

It ought to be possible to identify those factors or requirements which are either desirable or necessary to augur well for success in different circumstances. The resources in terms of equipment, facilities, funding and distribution networks, for example, needed by product-based firms will be different from service-based enterprises. This has implications for the kind of support and backup which each type of company requires if it is to prosper and this in turn will influence the types of structure and support mechanisms which a university, which wishes to become involved in its spin-out companies, will have to put in place to maximise the prospects for their success.

A key to this however, will be to characterise university/company interactions in terms of parameters through which identifying profiles can be drawn. It is postulated that in this way, the support mechanisms

147

appropriate to particular circumstances, as indicated by a particular 'profile', can be identified.

The Influences on a University Company

The experience of the practitioners in the present study finds support from previous research. Various researchers have drawn attention to the influence which the type of company has on the prospects for growth. Roberts (ibid.) for example, concluded that product-oriented firms are more likely to become high performers than are consultancies or research-based firms. Doutriaux (1987) noted differences in growth behaviour between technical service firms and manufacturing firms.

The extent to which the university's role and that of its staff affect new companies has also been assessed. In 1984, Segal Quince and Wicksteed showed that 17 per cent of the founders of 261 technology-based firms studied in the Cambridge area had come directly from Cambridge University. The university's attitude to commercialising of its research has been liberal, almost laissez-faire and that policy, it is claimed, has contributed to its effectiveness as an incubating organisation (Segal Quince and Wicksteed, 1990).

Smilor, Gibson and Dietrich (ibid.), analysed variables which acted as barriers or facilitators to spin-outs from the University of Texas. The university proved to be the most important organisation affecting company formation, 56 per cent of respondents in the study rated the university as being important or very important. In particular the university was valued as a source of personnel and of ideas. The role of the university in development was also recognised, 85 per cent indicating that it played a very important role as a source of personnel for continued company growth, as a resource of research expertise and ideas, as well as a source of scientific, engineering and business consultants. The authors concluded that the link between spin-out activity and university training in a technical field is strong whether the university educates or inspires the spin-out founder or simply recruits him or her to the area. In addition, the ability of these founders to continue their employment while also engaging in start-up activities has been an important characteristic of university spin-out company formation.

Doutriaux (1987) noted a negative effect on the growth and development of manufacturing firms of continued contracts with the university and also that manufacturing firms led by an entrepreneur still on the university payroll remained small and differed little from firms offering technical services whereas the other manufacturing firms in his sample became much larger. For the universities, the part-time professor / part-time businessman may make for efficient use of academic time and increase business contacts to the benefit of students and applied research. However, for the venture capitalist or the university wishing to maximise revenue from an invention or innovation, Doutriaux's research suggests that licensing out of the technology may be a wiser course, or that if a company is created to exploit it that it should be as independent as possible from the university.

The latter point lends support to the conclusion drawn from the survey (Chapter 8) of the difficulties which can occur when there is a failure to match the expectations of the various players involved in a start-up situation. In this case the requirements are to maximise the efficient use of academic time and the company's need to obtain a realistic return on its investment.

Links with the university and access to its services were significant factors which the smaller firms in Doutriaux's sample could not afford to have themselves.

A Swedish study by Klofsten and others compared the information requirements of university spin-out companies with non-university firms at start-up and later. The top three resources that were ranked by the university group to have internally, both at the start and for development, were technological know-how, personnel and marketing knowledge, Klofsten et al. (1988). Ongoing university linkages were also highly rated by university companies visited during the second interview programme to be described later (Chapter 13).

Taken together these studies reinforce the importance of the university's role on formation and development of companies and as sources of personnel, of ideas and research expertise. They emphasise the effect of the university's institutional involvement in its spin-out company or companies and the importance of the roles played by university staff as entrepreneurs, technical advisers or in executive posts.

Three Significant Parameters

The significant differences between the various universities, their companies and the diversity of relationships between them which were observed during the interview programme can all be explained in terms of the foregoing characteristics. This leads to the proposition that most situations can adequately be characterised by reducing the variables to three distinguishing parameters These three factors stand out as having a significant effect on the relative importance of many of the issues and decisions which are faced by university companies. They are:

- The nature and extent of the university's control of and involvement in the company;
- The nature and extent of involvement of university staff in the company;
- The products or services supplied by the company.

This approach builds upon and extends the definitions and taxonomies of Bullock (ibid.) and Lowe (ibid.) in the light of current practices. Furthermore the relationship between the parameters can usefully describe the 'profile' of interactions between a particular company and its parent university and this in turn can be used to determine the support mechanisms which need to be provided. The three parameters are represented in Figure 9.1 as a three-pointed star or Tripod, on to which a profile of a particular company can be drawn. Company/university relationships can be 'scored' against each of these parameters. The centre of the star represents the minimum and the outer extremity of each leg the maximum of each parameter.

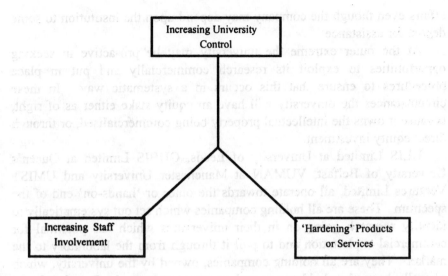

Figure 9.1 Parameters for constructing a profile for a university company

Parameter 1 - University Control

Nature and extent of the universities' control and involvement The role which universities play in the formation, subsequent development and control of new companies ranges from the disinterested or laissez-faire through permissive to the pro-active.

At one extreme, represented by the centre of the star in Figure 9.1, the university's role is passive or acquiescent, it may permit members of its staff to engage in entrepreneurial activity but the institution itself has no direct involvement in that activity. Such a laissez-faire attitude to its spin-outs as has been adopted, for example, by Cambridge University. This is typical of the 'hands-off' approach, in which the university has been content for its staff to engage in entrepreneurial activity without, as an institution, taking any direct part.

At a higher level of involvement the university may obtain payment from the company for use of university facilities, expertise or intellectual property, or it may receive a rental from the occupant of a Science Park but the university holds no equity in the company and takes no part in its affairs even though the company may depend upon the institution to some degree for assistance.

At the outer extreme the university may be pro-active in seeking opportunities to exploit its research commercially and put in place procedures to ensure that this occurs in a systematic way. In these circumstances the university will have an equity stake either as of right, because it owns the intellectual property being commercialised, or through direct equity investment.

ULIS Limited at University of Leeds, QUBIS Limited at Queen's University of Belfast, VUMAN at Manchester University and UMIST Ventures Limited, all operate towards the outer or 'hands-on' end of the spectrum. These are all holding companies which set out systematically to identify research going on in their universities which has potential for commercial exploitation and to pull it through from the laboratory to the market. They are all holding companies, owned by the university, which typically have shareholdings in and representation on the Boards of new companies which they form.

It is not necessary to have a holding company to have full university control of a company and many examples exist of wholly-owned university firms. The hands-on and hands-off models are at opposite ends of a continuum between which other equally valid arrangements exist.

Parameter 2 - University Staff Involvement

Involvement of university staff and the university's attitude Staff involvement may range from none at all, through part-time participation - for which permission may or may not have to be sought - to full-time involvement in a company. In the extreme, staff may be seconded from the university to work full-time for a commercial company.

University employees can be involved, with varying degrees of commitment as:

• Unpaid advisers;
• Paid consultants;
• University representatives on Boards of Directors;
• Investors/shareholders;
• Investors/Directors;
• Part-time employees;

- Full-time employees on 'secondment' from the university.

Examples of all of these are to be found throughout the spectrum of involvement.

Parameter 3 - Products or Services

Products or services offered by the companies At one extreme are those enterprises operating as general trading companies to provide university services such as accommodation, conference and catering facilities and university bookshops. They are not technology-based and are generally staffed by full-time employees of the company as distinct from those of the university. Often however the university will have a financial stake in such firms, if not complete ownership.

Moving outwards radially from the centre of Figure 9.1, one finds companies which carry out many of the activities normally associated with the Industrial Liaison Office of the university. The services offered may include testing, access to university laboratories or facilities, licensing of intellectual property and know-how and the provision of consultancy. Sometimes the consultancy will be provided by the company's own staff but more usually the company is acting as a broker for the university's academic staff. Some of these firms may also offer contract research through the university academic departments or specialised industrial units.

Specialised consultancies come next and typically consist of a group of specialists, marketing their services in a particular niche. These are usually scientists or engineers but include other disciplines such as management, economics and accountancy. Typically there will be a full-time core group supplemented where necessary with members of university academic staff. These companies have been describes as 'soft' in that they have relatively low fixed operating costs Bullock (ibid.).

Further up the scale come contract research and development companies. Typically these firms consist of specialists whose products may range from one-off machine design to special-purpose software but can encompass a wide range of technologies. Not untypically such firms are looking for a 'bread-and-butter' product and there is evidence from the survey that many started out with the intention of being manufacturers but for one reason or another have not yet developed a product to the stage where it can be marketed.

As companies move to the outer end of the spectrum their products 'harden'. Here are found those firms manufacturing products or providing services which require a greater commitment to a commercial approach and usually considerable investment in indirect as well as direct cost. In most respects these companies are indistinguishable from non-university based firms. Many have been formed opportunistically to exploit a particular invention or piece of university research. They are usually technology-based and typically the company's expertise is protected by patents or unique know-how. Some such companies have been very successful and become valuable assets, for example City University sold its technology arm for £20m (Financial Times, 1993b).

It should not be inferred that scoring either high or low on these scales is either good or bad. The purpose of the Tripod is simply to define a characteristic profile of a company and its relationships with its parent university so that better informed judgments can be made about the likely significance and relative importance of issues to be addressed, factors to be considered and infrastructure needed for the formation and operation of successful spin-out companies.

To enable these characteristics to be used more readily in practice they are defined more precisely below and allocated as degrees on a scale of one to five.

Summary of Degrees of Each Parameter

Nature and Extent of the University's Control and Involvement

Degree 1 - 'Hands-off' or laissez-faire The university permits members of staff to form companies or to become involved in them but takes no part as an institution either financially or otherwise in the companies' affairs. The university receives no income from the company.

Degree 2 - Passive commercial Allows companies to be formed, but takes no part in the process. Requires payment for use of university facilities used, e.g. space, equipment, laboratories, staff time.

Degree 3 - Encouraging Encourages the formation of companies but requires an equity stake or financial return by way of royalty or other income for use of university intellectual property and/or use of resources.

Degree 4 - Pro-active minor shareholder The university is pro-active in seeking opportunities to exploit its research through new company formation, has established procedures to encourage entrepreneurial activity but is content to be an active but minority shareholder.

Degree 5 - 'Hands-On' or pro-active significant shareholder The university is pro-active in seeking opportunities to exploit its research through new company formation. Has established procedures to encourage entrepreneurial activity. Provides back-up and support to fledgling companies and takes a significant or majority equity holding.

Involvement of University Staff

Degree 1 - Zero involvement No involvement of university staff in company affairs.

Degree 2 - Unpaid assistance University staff act as unpaid consultants or advisers.

Degree 3 - Paid assistance University staff act as paid consultants or advisers.

Degree 4 - Stakeholders University staff are shareholders and/or board members.

Degree 5 - Executive involvement University staff hold executive roles in the company, possibly on secondment.

Products or Services Offered by Companies

Degree 1 - General trading companies Companies which have been set up to provide university services such as conference facilities, catering, accommodation, travel agencies and company bookshops. No high-technology involvement.

Degree 2 - Industrial services Provision of university services such as consultancy and problem solving, access to laboratories and equipment, testing, analysis, licensing of intellectual property and know-how. Services required are usually referred to and provided by academic departments or university industrial units or centres.

Degree 3 - Soft companies Companies such as specialist consultancies comprising a full-time core group of specialists marketing their services in a well defined niche, sometimes augmented by the use of university staff when required. As well as engineers and scientists such groups may involve specialists in other fields e.g. management, economics, accountancy and law.

Degree 4 - Contract R&D companies Products may range from one-off machine design to special purpose software but can encompass a wide range of technologies. Often such companies are seeking, but have not yet found a 'bread and butter' product to sustain income and stimulate growth.

Degree 5 - Hard companies Companies manufacturing products or providing services which require a commitment to a commercial approach. Frequently high-technology-based and with a considerable investment in indirect as well as direct costs. Often these firms are indistinguishable from their non-university based competitors.

Example 1 - Typical Science Park Based Company

Of the 30 plus members of UNICO, about half are linked to science or technology parks which contain companies in which the university has no direct financial stake but in which university staff may be involved. Figure 9.2 shows a typical profile for such a company manufacturing technology-based products. Circumstances vary widely but, for example, the university's involvement may well be limited to that of landlord, although its staff may be employed as advisers to the company or as shareholders and directors. Products may range from soft to hard and are shown in Figure 9.2 as being hard.

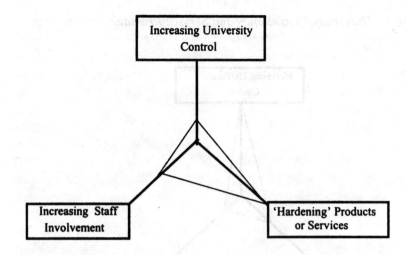

Figure 9.2 Profile of Science Park company using university consultants

Note that the score allocated to each of the three parameters enables a triangular profile to be constructed. The shape of this triangle will vary from case to case and it is this which identifies the relative importance of the parameters, uniquely, for each university/company combination.

Example 2 - University Consultancy and Service Provider

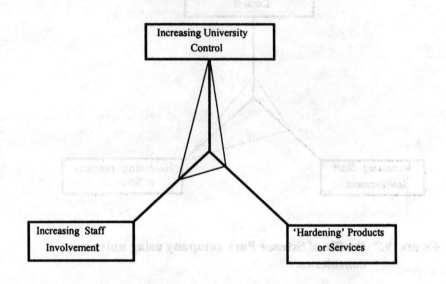

Figure 9.3 Profile of university owned company selling services/facilities

Perhaps the greatest number of wholly-owned university firms have been formed to exploit university consultancy and services, typified by the profile of Figure 9.3. These firms usually employ few university staff, other than on a casual basis, to whom consultancy and requests for help are referred.

Example 3 - The Holding Company Model

There is a discernible trend amongst UK universities to move towards a more managed approach to the exploitation of their technology. Figure 9.4 is a typical profile of a company formed by a university holding company making technology-based products. Typically the holding company will own in the region of (but rarely exactly equal to) 50 per cent of the equity, there will generally be some part-time staff involvement from the researchers most closely involved in the technology, usually as consultants and sometimes also as directors.

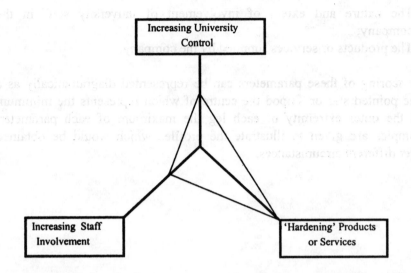

**Figure 9.4 Profile of partly university-owned company, using
university expertise to make technology-based products**

Summary

The characteristics of university spin-out companies have been
characterised by Bullock (ibid.) and others as occupying a spectrum
ranging from 'soft' to 'hard'. Published research indicates that the type of
company affects the prospects for growth and the survey carried out during
the present research has shown that the physical and financial requirements
also vary between different types of firm (Doutriaux, ibid.; Roberts, ibid.).

The importance of the university's role in the formation, development
and successful operation of university companies has also been
demonstrated (Doutriaux, op. cit.; Segal Quince and Wicksteed, ibid.;
Smilor et al., ibid.). This is supported by findings from the current survey
from which it is postulated that the main variations in university/company
interactions can be accommodated by defining three parameters.

• The nature and extent of the university's control of and involvement in
the company;

- The nature and extent of involvement of university staff in the company;
- The products or services supplied by the company.

The scoring of these parameters can be represented diagramatically as a three pointed star or Tripod the centre of which represents the minimum and the outer extremity of each leg the maximum of each parameter. Examples are given to illustrate the profiles which would be obtained under different circumstances.

10 Questions and Issues to be Addressed

Introduction

A number of questions must be asked and satisfactorily answered if university companies are to be formed with a reasonable chance of being successful. Some of these are driven purely by a priori considerations, others emerge from a distillation of the interview responses and subsequent discussions with academic entrepreneurs, a number arise from the literature and some are suggested by all of these considerations. Examples abound, with benefit of hindsight, of the effects of failure to ask these questions or perhaps to have answered them incorrectly.

It is implicit that the entrepreneur(s) or the university must have set up mechanisms and procedures both to ask the questions and to interpret the answers. The questions are mostly of the form How? What? or When? but in what follows they will be expressed in terms of the means to carry out the necessary tasks. The use of the terminology 'the means to' is to stress the necessity to put in place formal procedures to address the essential issues. Thus for example the 'means to assess commercial potential' implies that a methodology exists to ask and answer the question 'what is the commercial potential?' Only those matters which are of particular significance to the formation and operation of university companies, or which present particular difficulties for them, will be addressed.

Clearly some matters are more important in some circumstances than in others, for instance if the university is about to make a considerable financial investment it is more likely to be concerned about how to access funding than if it is simply being asked to support a low-cost consultancy practice. The relative importance of these tasks and activities to the university and the firm, will now be assessed for different university/company profiles using the 'tripod' profiling methodology described in the previous Chapter.

161

Essential Requirements

It is essential either to have or put in place the following:

- Means to assess commercial potential;
- Means to put in place entrepreneurs, product champions and mangers;
- Means to decide whether and when to form companies;
- Means to interact with university departments;
- Means to interact with the wider university system;
- Means to establish boards of directors;
- Means to recognise and deal with conflicts of interest;
- Means to access finance and funding;
- Means to assess performance;
- Means to reward inventors/ researchers;
- Means to provide support for embryonic companies.

Each of these will now be considered separately and their relative importance, as a university/company combination score moves along each the three Tripod parameters, will be indicated.

Means to Assess Commercial Potential

The decision to form a company is taken in the belief that there will be a market for the product or service to be provided. Whatever model is adopted someone or some group will have to take that decision. However, the consequences, for the university, of getting it wrong increase as it progresses from the hands-off towards the hands-on model i.e. from the apex of the Tripod to the outer end of the vertical arm (see Figure 9.1). The risk also increases with progression from the provision of university services through 'soft' companies such as consultancies, which usually have very low fixed costs, towards 'hard' high-technology product-based firms i.e. with progression along the right hand arm of the Tripod (Figure 9.1). The latter types of company are likely to require investment in full-time staff and frequently also in specialised equipment and facilities.

The mechanisms which the universities in the survey had put in place to decide on the commercial viability of ideas or innovations ranged from decisions based on fairly cursory assessments - near acts of faith in some cases - to those which relied on quite sophisticated market assessments and testing. For those inexperienced in market research, as many universities are when they approach it for the first time, it is all too easy to ask the wrong questions. The market research commissioned for one prospective biotechnology company asked about the size of the world market for synthetic peptides when a more pertinent question would have been how and when the potential new firm could expect to acquire a sufficient share of the market to become profitable.

It is at this stage that the complex and iterative nature of the innovation process is most apparent. There is also a tendency to confuse needs, as perceived by researchers and inventors, with wants which can be translated into sales. For the inexperienced, confusion can easily occur between genuine market demand and technology-led ideas which may well have the potential to stimulate a potential market but which frequently require further development, design for manufacture, innovative marketing, or all three.

Timing is also critical and the ingredients for success include not only the determination of perceived market needs but the means and ability to turn that demand into actual sales in a reasonable time scale. This is particularly true where a large investment in equipment or other fixed costs is envisaged as this usually implies that fairly high sales revenue is required to reach a break-even point before the company can be profitable. Until this is achieved the requirements for working capital have to be met from other sources.

Getting an assessment of likely market demand is most critical as university involvement increases and as the company moves towards the product-based or 'harder' end of the spectrum of product types. i.e. as the profile moves higher up the vertical leg and further out along the right hand leg of the Tripod of Figure 9.1. The iterative nature of the innovation process should not be overlooked and special expertise may be needed to avoid costly mistakes by making market assessments at too early a stage of product development.

Means to put in place Entrepreneurs, Product Champions and Managers

The need for procedures to address this issue is very apparent from the literature and was supported by the survey findings (Harvey, 1994; Oakey, et al., 1988; Pickles and O'Farrell, 1987; Roberts, 1991; Samsom and Gurdon, 1990; Sonderstrom et al., 1986).

There was widespread agreement amongst practitioners, both from the university administrations and amongst academics, about the need to have both product champions and entrepreneurs. Although there were difficulties with precise definitions, in most companies it was possible to identify the entrepreneurs and product champions in hindsight. Only rarely, however, were both roles fulfilled by one individual. More often than not in university companies the product champion emerged from the ranks of the researchers. Very often these people saw themselves also as entrepreneurs but there are many examples of embryonic firms failing to succeed because the inventors were left in charge.

There is considerable support for the view that a mix of skills is required and also that often the original entrepreneur is not always the most appropriate person to lead the company as it matures (Samson and Gurdon, ibid.). Frequently it has been possible to recruit entrepreneurial managers and ideally these are complemented where necessary by a board of directors with marketing and financial ability to provide the combination of expertise needed to be successful. Although guidance on the characteristics of those most likely to be successful is available from the literature (e.g. Harvey, 1994; Roberts, ibid.; Pickles and O'Farrell, ibid.) none of the universities had consciously addressed these findings in their search for entrepreneurs or managers. In practical terms it would be difficult under current fair employment legislation to justify an appointment on the basis of, say, a candidate's sibling position or father's occupation.

There was broad agreement amongst those interviewed that full-time managers were preferable and that it was undesirable for management to be left to part-time academics. Sometimes the manager may also be the entrepreneur or the product champion but it is comparatively rare to find a single individual who can successfully combine these roles.

The need to put in place entrepreneurs, product champions and managers is important in all cases but increases with progress towards the manufacture of 'harder' products.

Means to Decide Whether and When to Form Companies

Although only a few universities referred explicitly during the interviews to the need to maximise (or optimise) their overall funding, more recently this seems to have received increased attention. From the survey it is clear the alternatives to company formation were not always evaluated and some universities admitted to having been persuaded to embark on the new company route by enthusiastic, often senior, academics. In hindsight a number of companies were formed unnecessarily or too soon. A degree of naïveté existed amongst many academics that somehow forming a company would automatically be beneficial to the prospects of commercialising their ideas. Other means of exploiting technology such as licensing, staff exchanges and collaborative research are sometimes more appropriate and the discipline of comparing and recording the reasons for the choices made is well worthwhile, for example see Doutriaux (1987) and Oloffson and Wahlbin (1984).

Timing is also more important than is generally realised. Considerations relate inter-alia to the state of the national and regional economies, the likelihood of a change of government, the available grants regime, the activities of competitors and the state of competing technologies.

Choosing the most appropriate method and optimum timing to exploit university technology is of particular importance where decisions are to be taken to commit to a product-based business and especially so where substantial financial investment is required.

Means to Reconcile Objectives and Expectations

Although it is clear from the literature that there are likely to be divergent motivations for forming spin-out firms, the main impetus to address this question arose from the interviews. Many instances were related in which differing expectations of the university administration and those of would-be academic entrepreneurs were not apparent at company formation stage. This usually occurred where the objectives of the company had not been formally stated and agreed by all concerned. The difficulty of achieving this consensus should not be under-estimated. Several cases were recounted of researchers with 'hidden agendas' which if brought into the open could often have been accommodated in the initial business plan but which emerging later caused disaffection and frustration. Different aspirations need not be mutually exclusive but can be damaging where they go unrecognised. The process of agreeing company aims is vital where both the university as an institution and members of its staff are deeply involved in an enterprise.

It is equally important that the expectations of third parties are understood and agreed upon. The respective contributions, both financial and in kind, of all the equity partners should be formally agreed in writing and incorporated in the Articles of Association of the company and where appropriate as part of a formal Shareholders' Agreement. Practitioners with experience of equity sharing ventures stressed the need to safeguard the university's position in cases where it was to be a minor shareholder and therefore at risk of being oppressed by other significant shareholders. Amongst the most important issues to be included is the university's or holding company's right to appoint a director to the Board of the company, the provision of a mandatory dividend and agreement about wind-up or sale of the company.

Attention to this issue is most important when there is likely to be a high involvement of university staff in the proposed company i.e. as the profile moves outwards on the left hand Tripod arm, Figure 9.1. It is also of consequence when the university envisages making a significant financial investment for a minority shareholding.

Means to Interact with Academic Departments

Again the importance of this issue emerged during interviews with university practitioners and entrepreneurs. Although there were many instances of good collaboration between companies and academic departments and their staff there were also some horror stories. The latter generally related to the sharing of resources including staff time, space, equipment and facilities. Instances were quoted of changes in attitude of university departments to erstwhile colleagues after companies had been formed. Examples were recounted of companies suddenly being charged for occasional photocopies and use of, often otherwise unused, equipment and space. In some cases this resulted in counter-charging for the use by university staff of company equipment. There is no reason why charges should not be made to companies and universities for legitimate use of each other's resources but the relationships which worked best were those where informal quid pro quo arrangements prevailed in which the parties recognised mutual benefit. In some cases difficulties could be traced to a lack of identification with the company and its aims by other academic researchers. It is easy to dismiss this, as some company staff did, as jealousy or to the fact that some people were 'difficult'. There was general agreement that it is desirable to try to anticipate and avoid these problems at an early stage, particularly where there is a need for a sharing of resources, including staff time. This however requires a conscious effort to do.

To achieve good collaboration often required a good deal of effort, especially from companies located in university departments who felt a need to demonstrate that they were not cuckoos in the academic nest. Some companies had incorporated departmental representation on their Boards to aid with identification, others had gone to considerable lengths to sponsor studentships and research projects or to donate equipment to academic departments. There were examples of initially difficult relationships being transformed as a result of such efforts.

This issue grows in importance with increased involvement in the company of university staff and departments i.e. as the profile progresses outwards along the left hand arm of the Tripod.

Means to Interact with the Wider University System

The importance of this issue and how it is dealt with depends on the involvement of the university as an institution in the company's affairs. Where there is formal institutional representation on the Board of a company or on the Board of a holding company, as distinct from simply the participation of research staff, the problem may be minimised. It appeared to matter most when issues relating to ownership and exploitation of intellectual property were involved, where there was to be financial investment by the university and on matters relating to the use of university resources and personnel. The general ethos of the university and its approach to its companies was an important factor as was support from top management of the university, see Chapter 8.

Means to interact with the wider university system can be vital to success and become most important as the degree of university involvement and stake in the company grows.

Means to Establish Boards of Directors

As companies move along the scale from soft to hard it becomes increasingly important to have a proper balance of boardroom skills. By their nature, university spin-outs tend to be well provided with technical expertise but the innovation process and companies engaged in it need more than this. Much has already been said about the need for an entrepreneur and product champion, one or other of whom may be the chief executive but as companies develop, a range of complementary skills becomes necessary (Benedetti, 1987; Morita, 1992; Rothwell and Zegveld, 1981). In particular, financial expertise and marketing ability will have to be provided. These can often be provided by a university holding company but they may also be obtained from non-executive directors or the participation of representatives from industry and business (Cartin, 1992).

A widely experienced Board of Directors becomes more important as companies progress from soft to hard product-based firms.

Means to Recognise and Deal with Potential Conflicts of Interest

The survey results confirmed the prevalence of conflicts of interest but a surprising insouciance on the part of those experiencing it and a confidence that it could be dealt with without detriment either to the companies' or the universities' interests. Although acknowledging its presence it was simply not regarded as being a serious problem by those interviewed. This does not mean however that it can be ignored and it is possible that others who were not interviewed, for example academic colleagues of the entrepreneurs, might view the position differently.

This is obviously of greatest importance in those companies where university staff are involved. Perhaps the greatest danger is that problems are either not recognised or, as was shown during the interviews, they are rationalised by those involved as being unimportant or related to issues which they can always reconcile.

A recent and potentially more difficult problem faced by the UK universities will be how to deal with the tension, referred to earlier, between the pressure to retain high research assessment ratings on the one hand and to commercialise their research on the other. This problem was being anticipated during the latter stages of the survey but no university had at that time had to grapple with it seriously.

Although one serious conflict of interest is one too many and all universities and their companies must be vigilant, this problem is potentially at its greatest where there is high involvement of university staff.

Means to Access Finance and Funding

This is an a priori consideration and although the need to access funding does not apply uniquely to university companies, there are special factors, not least of which is risk, associated with the funding of NTBFs. There is also a good deal of guidance on the appropriateness of alternative sources in the literature, for example Cartin, 1992; Mason and Harrison, 1994; Philpott, 1994 and Standeven, 1993.

The most difficult type of funding for universities to obtain appears to be that required, typically a few tens of thousands of pounds, to develop an idea or to produce a prototype to the point where its viability can be demonstrated. There is a role for government pump-priming in this area and indeed the DTI-funded SMART award scheme has been particularly useful to a number of university firms in providing this initial funding. Regional development agencies have also been able to assist in some cases. Some universities have provided this type of pump-priming from their private resources. The University of Newcastle, for example, formed its own venture-capital fund to provide injections ranging from £2,000 to £50,000 to help exploit its research (Financial Times, 1993a).

Once this embryonic stage has been passed there is a proliferation of funding sources, including private investors, existing companies, venture capital firms and even stock-market flotation. See for example the flotation of Aromascan by UMIST Ventures Limited (Financial Times, 10 July 1994). Examples of all of these have been successfully used by UK university firms.

It is perhaps in this area that a university holding company has most to contribute to its offspring. Expertise, contacts and, most importantly, credibility with potential investors, can be built up to enable new companies to obtain both first-stage and subsequent development finance.

The importance of having means to access funding increases with the level of investment required but in general it is more likely to be of concern where the university has a high degree of involvement with the company.

Means for Assessing Performance

It is a sine qua non for performance assessment that objectives must be unambiguously set. Whatever these are, however, the financial realities are that for the company to survive, income must eventually exceed expenditure. In the commercial world income is expected to come from customers as sales revenue but some university companies have been in receipt of grants, for R&D for example and from government agencies. There is a real danger that grant income is confused with sales income and this is particularly likely to happen in a university setting where it is common for most research income to come in the form of grants as distinct from earnings.

On the expenditure side many companies receive hidden subsidies from their universities, such as free or subsidised accommodation and either a reduction or complete waiver of 'overhead' charges. Notwithstanding any non-financial objectives which the company may have, the financial reality should be faced and there is a need in all companies for proper financial controls.

Although the need for financial controls is paramount to ensure economic survival there is also a need to monitor the company's non-financial objectives. Most companies have objectives related to turnover, market share and product development. It is perfectly legitimate for example that a firm should have amongst its aims to increase research income for the university or to employ its graduates, provided that these are compatible with the other market-orientated objectives (see Roberts, 1991). The monitoring of performance against all of a company's objectives is part of the role of management and the Board.

Strictly speaking this issue should be important in all cases, if only for the discipline which it brings. It is however at its most important for a university which has a significant financial stake in a company.

Means of Rewarding Inventors/Researchers

Whatever model was adopted for new company formation, there was widespread recognition amongst the interviewees that the researchers whose work was to be exploited should be 'rewarded' in some way. The survey confirmed the findings of much of the literature that the motivations of researchers were not always, or even mostly, financial. Recognition of their work and seeing it exploited commercially was often valued more that pecuniary gain. The situation is complicated by the fact that the reward system in academia is linked more to the publishing of academic papers and dissemination of knowledge than to its commercial exploitation. Academic promotion is unlikely to follow solely from successful technology-transfer or commercial success.

Some researchers interviewed were quite happy simply to be named as directors of companies, others sought consultancy fees. A fairly common practice involved the provision of equity in the company either at no cost or at a discount in exchange for the use of intellectual property or expertise.[1] It should not be overlooked that intellectual property need not be patented or formally protected to be valuable. Unique expertise or 'know-how' can often be commercially exploited. A very few sought control over 'their' companies but the overwhelming experience has been that academic researchers rarely make good managers and that they are best employed as technical directors or consultants. In any case, it is necessary to reach an amicable agreement in order to ensure continued and mutually beneficial collaboration.

This issue is likely to most important where there is a high staff involvement in a project, although there are instances of payments or share offers being made to academics who play no direct part in a company's operations.

Means to Support Embryonic Companies

Most, but not all, of the universities surveyed recognised the need for early support to embryonic firms. Some supplied practical help and services from central university administration and accounts departments, although it should be noted that university accounting systems are usually quite unsuited to providing the kind of information required by small firms. Better provided for were those firms supported by university-owned holding companies. The provision of simple management accounts, and not having to deal with sales and purchases, wages and VAT returns all enable the management of the fledgling enterprise to concentrate on establishing itself and its products or services without becoming side-tracked with day-to-day administration. Sometimes these services are provided free, or nearly so, by the holding company but most universities recognised that eventually all companies should become self-sufficient. Nevertheless, some examples were noted where quite well established firms were still dependent on support from a holding company, but on a fee paying basis. Examples of the types of services which can be supplied are contained in the case-study described in Chapter 12.

The provision of support services is most important where the university has a significant or major financial interest. Frequently these services are provided via a holding company.

Summary

A number of questions and issues have been identified from the interviews, from the literature and a priori. The importance of these issues varies depending on the characteristic Tripod 'profile' of the company/university relationship. The matrix in Table 10.1 summarises those requirements which are of greatest importance at the outer extreme of each parameter i.e. maximum university control, maximum staff involvement and the production of 'hard' products.

Table 10.1 Relative importance of requirements at extremes of each parameter

Requirement	Increasing University Control	Increasing Staff Involvement	'Hardening' of Products
Assess commercial potential	High		High
Decide whether and when to form companies	High	High	
Put in place entrepreneurs and product champions	High		High
Appoint full-time managers	High		High
Reconcile objectives and expectations		High	
Interact with academic departments		High	
Interact with the wider university system	High		
Establish Boards of Directors	High		High
Recognise and deal with potential conflicts of interest		High	
Access finance and funding	High		High
Assess performance	High		
Reward inventors and researchers		High	
Support embryonic companies	High		

Note

1. Even though intellectual property (IP) in the UK belongs to the university, as employer, rather than to the researchers, the reality is that to exploit the IP it is usually necessary for the academic staff who understand the technologies concerned to become involved. Their cooperation is necessary for any type of commercialisation to occur.

11 Emerging Structures

Pragmatic Versus Systematic Approaches

Although the interviews and discussions revealed great diversity of approach by universities to new company formation, most had embarked on this course in a pragmatic way - either to market their services and facilities more effectively or in response to opportunities to commercialise particular technological breakthroughs or inventions. Fourteen of the initial 16 universities visited came into this category and only two, had attempted systematically to exploit their intellectual property by forming new companies. This ad-hoc approach is the most common amongst UK universities - the vast majority of the 30 plus university companies in UNICO - were formed in that way. Results have been mixed but there is no doubt that some very successful companies have been formed in this opportunistic way, examples include Oxford Instruments and City University's gas sensor company. Although widespread, this is a somewhat haphazard approach to exploitation which relies heavily on ideas being brought to the attention of the university by someone with entrepreneurial motivation and the vision to see the commercial opportunities. When companies are formed on an occasional basis there is usually no supporting infrastructure to assist with the process and, if this occurs only infrequently, the lessons have to be relearned each time.

By contrast, an increasing number of universities are beginning to adopt more proactive and systematic procedures to commercialise their research and a handful have formed some variant of holding company to do so. Universities adopting this approach typically carry out regular trawls or otherwise keep a continuous audit of ongoing research which may have commercial potential. Companies formed under this type of regime are unlikely to occur fortuitously. The decision to form a company will have been reached only after consideration of a variety of factors not least of which will have been alternative means of exploiting the technology, such as licensing.

It is was not possible during our research to identify best practice. Further research into the relative effectiveness of this approach versus the more widespread pragmatism would require a much larger sample of companies and there may not yet be enough mature companies in the 'systematic' grouping. An a priori case could, however, be made for the superiority of the systematic approach. Firstly, because it is reasonable to assume that fewer opportunities for exploitation will be 'missed' if pro-active monitoring of ongoing R&D occurs. Secondly, because of the existence of an infrastructure to assist with activities such as market assessment, preparation of business plans, search for potential partners and funding as well as the procedural and legal steps necessary for company formation. Thirdly, for the ongoing support and back-up which can be provided to embryonic companies from a holding company of the kind to be described. The foregoing presupposes however that universities should be engaged in this type of activity in the first place. There are those who would argue otherwise, so the effect of the null hypothesis i.e. of not becoming involved in entrepreneurial activity, should also be evaluated.

The Holding Company Model

Two universities included in the original visit programme had formed holding companies. AURIS Limited at Aberdeen and QUBIS Limited at Queen's in Belfast are examples of what may well be the next generation of university-linked companies in the UK. Eight UK universities known to have adopted this approach and which together had formed about 90 companies by 1995, are alphabetically[1]:

- AURIS Limited, University of Aberdeen;
- QUBIS Limited, Queen's University in Belfast;
- SUBSL, Salford University Business Services Limited;
- UCL Ventures Limited, University College London;
- ULIS Limited, University of Leeds;
- UMIST Ventures Limited, UMIST in Manchester;.
- VUMAN, University of Manchester;
- Zeton Limited, University of Nottingham.

Having identified the trend towards the holding company model, a second series of visits was made to include data from UCL, London; ULIS,

Leeds; UMIST and VUMAN, Manchester. The University of Twente in the Netherlands, which has formal procedures for its TOPS and TOS programmes was visited and Trinity College Dublin which has adopted a systematic approach but which has a somewhat unusual structure was revisited.

A detailed description of the QUBIS Limited group of companies is included in Chapter 12. Many of the features described in that case-study exist in the other holding companies although there are some differences in approach. One holding company, for example, has arranged that the intellectual property rights (IPR) arising from research are vested not in the university but in the researchers. This, it is claimed, creates a less threatening and more encouraging atmosphere for researchers to exploit their research through the company.

Whereas most of the holding companies form new companies at the earliest opportunity, several have separate operating divisions as well as having minority stakes in their subsidiary or associated companies. One of those included in our survey waits until the operations are well established and with annual turnover in the region of £250k before formal company formation. In the interim the 'companies' exist as separate operating divisions of the holding company. Another unusual facet is that when the companies are in due course split-off, the equity stake is held not by the holding company but by the university. The proportion of equity held also varies. Most holding companies are happy to hold minority shares in their spin-outs and in some cases prefer to do so, although some other universities are less willing to relinquish control.

Sources of funding differ. UMIST Ventures has successfully floated two of its companies, Tepnel Diagnostics and Aromascan, on the London stock-market and raised many £ millions for these ventures (Sunday Telegraph, 5 June 1994). UCL Ventures has close links with Natwest Bank and has turned to informal capital through 'business angels' to match funding obtained from DTI SMART awards for its companies. QUBIS Limited has pioneered the concept of corporate venturing in which it shares equity in its companies with established firms. This has the advantage that the external companies bring not only funding to the deal but also marketing and commercial expertise which complements the technical innovation provided by the university researchers (Cartin , 1992).

All the holding companies provide support in various ways. including early stage support, such as the preparation of business plans, market research, financial accounting and project evaluation for potential academic entrepreneurs. In all cases the university holding companies have Board representation on their companies. The chief executive of one university holding company, for example, is chairman of each of the subsidiary firms.

As well as their roles in new company formation VUMAN and ULIS both carry out some of the functions normally dealt with by university industrial liaison departments. This includes dealing with externally funded research contracts for academic departments, consultancy assignments for university staff and the licensing of university IPR and know-how. AURIS works very closely with the University's External Funding and Industrial Services Office (EFISO) to the extent that they share the same building and operate under the same name and are attempting to market themselves to the outside world and internally as a single entity.

QUBIS Limited, in contrast, concentrates exclusively on new company formation but works very closely with the industrial liaison office of the University with which, until recently, it shared offices and a common acronym.

A Systemic View

Although all of these holding companies differed in detail, they had in common a view that the exploitation of research should be a managed activity as distinct from something which happens by chance. To understand them better they can each usefully be regarded as university-owned systems for exploiting research through new company formation (Blair, 1990b).

A systemic approach is particularly appropriate because university spin-out companies - their creation, operation, management and control and the complex web of relationships involving inter-alia researchers, university departments, administration, company staff, and external partners and funders - involve situations in which human perceptions, expectations and actions often dominate. Because the individuals and groupings involved are likely to have differing motivations and values, situations can occur where even the interpretation of events becomes problematic. The ambiguous nature of replies received about how to measure the success of companies was symptomatic of just this kind of difficulty. Although beyond the scope of what will be attempted here a 'soft systems' approach to understanding the structure and nature of the interactions would be an appropriate tool for further research[2].

The Core of the System

Figure 11.1 shows the holding company and its various activities at the core of the system. Although a system boundary can be drawn around the holding company which usually has a separate legal identity from the university, for practical purposes its operations are nevertheless part of and within the wider university system. The whole is of course influenced and perhaps constrained by the external environment.

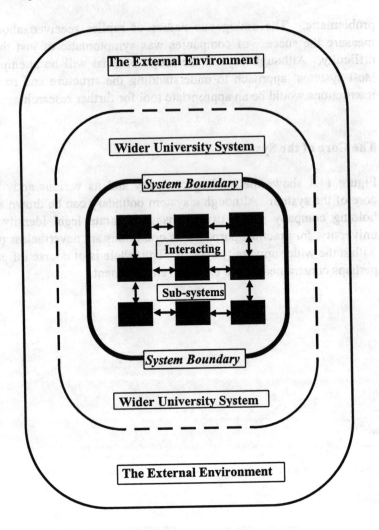

Figure 11.1 Outline system diagram

At the core of the system is the holding company. The various activities it carries out and the services it provides are represented in Figure 11.1 as interacting sub-systems. These varied from university to university but typically included the monitoring of ongoing R&D for commercial potential, market assessment for potential new products, production of business plans, searching for funding and equity sharing partners, managing interaction with university departments, managing individual companies and providing support services for embryonic firms.

The fact that there may be a long list of activities (sub-systems) does not imply that the staffing of the holding company need be large and in fact those visited employed between four and eight staff. Holding company staff may engage in different tasks at different times and for activities such as the evaluation of market demand, university staff, external personnel, or consultants are often involved. The staff of the holding companies were almost all employees of the companies as distinct from the universities. As a result they had a degree of independence from the universities although, without exception, the chief executives of these companies stressed the importance of maintaining very close ties and working relationships with their parent university administrations as well as with departmental academic staff. Expressing the model in systems terms, the overall system and its sub-systems have inputs which are transformed into a range of outputs. The inputs include for example - ideas, innovations, expertise, physical resources and funding. The outputs may include business plans, embryonic and growing companies, intellectual property and an income stream for the shareholders and for the university as owner of the system. The sub-systems incorporate mechanisms for decision-making and performance monitoring. An important reason for viewing these activities in this way is to emphasise that they are all to some degree inter-related, that the system is holistic and that it has emergent properties i.e. the whole is greater than the sum of its parts.

The Influence of the Wider University System

The university system comprises the academic faculties, schools and departments as well as the central administration. It may also include specialist industrial centres or units which are involved in technology-transfer, by other means, to industry. The impact of the wider university system is firstly to authorise or legitimise the system for technology-transfer through new company formation and possibly also to impose constraints upon its activities. It usually imposes rules or procedures to protect the university's interests. For example a number of institutions have laid down regulations for the licensing and use by spin-outs of university intellectual property, others have formal procedures for using and paying of staff and other resources used by companies.

The discipline of viewing the various activities in systemic terms forces consideration of how they interact both within the holding company and with the university. For example, in one holding company a decision to form a company based on laser technology developed by an academic professor almost foundered because of opposition by his department head to the use of laboratory space and equipment. The trouble arose because, although the head of department had approved the project in principle, he was dissatisfied with the amounts offered by the company as payment for the use of necessary departmental facilities. Even though the professor closest to the project understood the position, it became apparent that because his department head had not grasped the difference between sales revenue and profitability, the department head assumed that all the company's sales income was available for distribution.

The lesson learned by the holding company was of the danger of assuming that even eminent academics understood the basic economic facts of business. The procedures introduced thereafter ensured that not only were departmental staff kept informed but that the heads of departments fully understood the implications of new companies arising from their departmental research. An additional activity was consciously added to those of the holding company.

Another holding company has set up a formal procedure to avoid conflicts of interest between academic entrepreneurs and the university. Frequently embryonic companies require ongoing research or product development from the university and usually this is provided by the academic researcher(s) involved in the new company. This has the potential to create tension between the researcher's role as academic staff member - with a responsibility to obtain economic fees for the university - and the requirement to get the best possible deal for his/her company. The procedure established by the holding company is that all contractual matters between spin-out companies and the institution are negotiated, with the university's industrial liaison office, rather than with the researchers involved or their academic departments.

The impact which the wider university system has on the operations of any spin-out company are greatly influenced by the extent both of university involvement - being greatest at the 'hands-on' end of the spectrum - and of the extent of staff involvement in company affairs. All of the UK holding companies visited were operating with a high degree of university involvement.

Effect of the External Environment

Any system for exploiting technology is bound to be influenced or disturbed by the external environment in which it operates. Few universities, even those which operated holding companies admitted to thinking about these external factors when setting up their companies, yet in hindsight they were quite influential on the success of their ventures.

Surprisingly there was little reference during the interviews to the current economic climate although it obviously had an effect. Regional factors and government policies had more direct impact. For example, the availability of government grants and assistance to new start-ups make it easier to obtain pump-priming finance in some areas than in others. On the other hand the scarcity of existing technology-based firms in development regions makes it more difficult to find local partners for joint ventures. The helpfulness or otherwise of local government economic development agencies was a factor. So too was local competition faced by university firms. Relatively mundane matters such as a crowded city centre and high rents affected some universities more than others. The matters to consider will differ from time to time and geographically.

Summary

Although most UK universities have responded opportunistically in forming companies to exploit their research, an increasing number are now adopting a more systematic and proactive approach. Rather than reacting to approaches from academic staff or departments to form companies, these universities have put in place systems to identify and commercialise research with potential for commercial exploitation. A handful of UK universities have pioneered this approach by forming holding companies, with primary emphasis on new company formation, and others are beginning to emulate it. The differences in the models adopted include the equity stakes retained by the holding company, their preferred sources of funding, their equity sharing partners, and the range of services and support which they offer to their subsidiary and related companies. It is helpful to view this model in systemic terms. In this way the necessary support mechanisms and procedures which have to put in place can be identified as being interacting sub-systems. The influence of the wider university system and the external environment can also be assessed.

Notes

1. There is a larger group of UK universities which have formed companies primarily to market their services to industry including R&D and consultancy but those listed above are those known to be primarily concerned with new company formation. All are using a holding company and all attempt systematically to exploit their university's technology through new company formation.

2. Soft systems analysis differs from more familiar hard systems methods which attempt to optimise (usually) agreed and objective solutions. It is close to 'action research', in that the analyst is attempting not only to understand the situation but to change it for the better (Open University, 1984; Checkland, 1981; Ackoff, 1974).

12 Case-Study - The QUBIS Group

Introduction

This Chapter is a detailed case-study of commercial spin-out activity from the Queen's University of Belfast during the period 1984 to 1996 promoted by a university company set up specifically for this purpose. The purpose of the exercise is to examine in detail one example of the holding company model.

Following the systemic approach described in Chapter 11, the holding company QUBIS Limited and its associated companies will be regarded not as separate entities but as elements of a system for the commercial exploitation of university technology. It is necessary, therefore, to explain the environment and the context in which this system was formed and within which it operates.

Context/Environment

To set the case-study in perspective it is necessary to appreciate the environment in which QUBIS Limited and its companies operate. Northern Ireland is the most westerly region of the United Kingdom. It has a population of just over 1.6 million, increasing by about 0.3 per cent each year and one of the highest unemployment rates of any UK region. There is a heavy dependence on services which now accounts for 72 per cent of those employed, a large proportion of whom work in the public sector. Only about 100,000 people are employed in manufacturing industry and there is still quite a high reliance on agriculture and the traditional industries of textiles, ship building, clothing and tobacco many of whose products are at a mature stage of their life cycles (DFP, 1989; NIEC, 1990). A priority, therefore, is to broaden the industrial base to include newer and emerging technologies and ideally to create and assist knowledge-based, high added-value companies, not handicapped by location or transportation costs.

The situation provides a mixture of opportunities for and constraints upon technology-transfer. The supportive attitude of government and emphasis on job creation are positive factors, whereas remoteness from markets and the limited industrial base place limitations on the extent of involvement with local companies. The stated aims of QUBIS Limited coincided exactly with government's priorities of encouraging the creation of more high-tech companies. Government agencies assisted, therefore, at an early stage by making funds available on a £ for £ basis, for initial evaluation of potential projects and later to help attract joint-venture partners.

The Wider University System

Queen's is the older of two universities in Northern Ireland. Established in 1845 it has a student population in excess of 20,000, the majority of whom come from Northern Ireland. These comprise 9,000 full-time and 1,100 part-time undergraduates, 1,800 full-time and 1,800 part-time UK postgraduates and, 450 overseas students and 6,600 students on continuing education courses. Like all UK universities, Queen's has been under pressure from government to obtain an ever increasing proportion of its income from non-government sources.

The University does not have an enclosed campus but occupies a number of major university buildings centred on the original University building built in 1845 by Sir Charles Lanyon. A great many smaller houses and premises have been acquired over the years so that the Queen's campus dominates and encompasses an area radiating about half a mile from the Lanyon Building. There are additionally a number of outposts, at three teaching hospitals and the Faculty of Agricultural and Food Science which are each between one and two miles away and on the shores of Strangford Lough there is a Marine Biology Centre.

The wider University system includes its academic community, managerial structure and administration but of particular relevance for technology-transfer are the other mechanisms and units for dealing with industry. Queen's has a long tradition of working with industry and provides a wide range of services including consultancy, R&D, problem solving, licensing and access for industry to University facilities. There are also a number of specialised industrialised units which together employ about 90 people solely on industrial activities. It was to extend this range of services and to enable it to enter into equity sharing joint ventures that the University decided to form QUBIS Limited. The Company's activities complement these longer established mechanisms for technology-transfer and exploitation of intellectual property.

The ways in which QUBIS Limited and its companies operate had to be consistent with the procedures and practices of other University groups. For example, it was important, from the outset, to ensure that the new companies did not compete with the academic departments or undercut the fees charged by the University for its industrial services. Where the companies make use of University resources or intellectual property they pay realistic charges. The University Vice-Chancellor, two Pro-Vice-Chancellors and Bursar are members of the QUBIS Limited Board; there is therefore a degree of monitoring by the University of the Company's activity.

QUBIS Limited - The Parent Company

Background and Approach

QUBIS Limited was formed in 1984 to commercialise the research and development being carried out at the University by identifying R&D with commercial potential and 'pulling it through' to the market. It does this through the formation of new companies, ideally sharing ownership with established companies whose commercial skills or market position complement the University's technical and innovative ability. This type of corporate venturing has become one of the distinguishing features of QUBIS Limited. Accordingly, although the possibility of technology-push is not ruled out, most projects are market led to the extent that an established company believes that a market either exists or can be created for the product.

The vital ingredients for successful venturing are believed to be:

- Strong evidence of a market or a potential market;
- The presence of a 'missionary' i.e. someone who will regard it as his/her mission to make the venture succeed;
- Securing the best possible management for the project;
- Appropriate partners, the most successful QUBIS Limited ventures have experienced industrial or commercial partners (Cartin, 1995).

Shareholding Policy

As a result of its corporate venturing policy, QUBIS Limited is often content to be a minority shareholder, allowing the major investor(s) to take first responsibility for ensuring the success of the venture. QUBIS Limited takes steps to protect its position through the Articles of Association and, where appropriate, through Shareholders' Agreements which, for example, enshrine its rights to appoint directors and to receive mandatory dividends.

It is also policy to give incentives and rewards to University researchers by offering shareholdings. As a matter of policy and an indication of serious intent, a minimum investment of £1,000 is sought from researchers who wish to become shareholders in QUBIS Limited companies. Often, however, 'equity for expertise' arrangements are made and the shareholdings obtained by staff are often greater than their cash investment would otherwise entitle them to. Members of University staff have shareholdings in all of the QUBIS Limited companies.

With a given portfolio of businesses, some partners, the management or other interested partners may wish to buy some or all of the QUBIS Limited stake. However, QUBIS shareholdings in its joint ventures are structured for long term relationships (Cartin, 1995).

Staffing, Directors and Management

QUBIS Limited has an authorised share capital of £500,000 of which £381,000 has been issued; it is wholly owned by the University but has limited liability status, employs its own staff, three full-time and one part-time and operates with a great deal of autonomy.

Half of the QUBIS Limited directors are drawn from industry and the local business community. With hindsight this was a wise decision. Not only have the external directors been able to offer sound advice, they have brought a sense of commercial realism to the Board and added greatly to the credibility of the Company in its dealings with potential joint venture partners and with government. As directors and frequently in the role of Company Chairman, these experienced industrialists offer a lifetime's expertise to embryonic spin-out firms free of charge. This is believed to be one of the reasons why none of the companies has failed so far. The Board meets quarterly and day-to-day management of the Company is via a Management Committee which meets monthly.

Objectives and Targets

The objective of QUBIS Limited is to establish a stream of dividend or other income from its investments. In building a portfolio of successful investments in spin-outs from the University's technology there are also several by-products:

- Economic development through new company formation and creation of employment, especially for graduates;
- Transfer of knowledge, mainly technology to local and national industry;
- Enhanced University profile which highlights commercially relevant research.

The immediate targets of QUBIS Limited are:

- To establish one or two new business ventures each year, with outside partners where possible, and to encourage academic staff to take a stake in each new enterprise;
- To maintain and grow the existing businesses both in terms of employment levels and profitability year by year until a sustainable dividend can be paid to QUBIS Limited's shareholder, The Queen's University of Belfast.

Service ventures are expected to become profitable at an early date, year two being the target. More capital intensive, or R&D led ventures, are expected to make profits in years three to five (Cartin, 1995).

Location

QUBIS Limited itself is located on the University Campus but in a separate building, deliberately chosen to be neutral in that it is not identified with any particular academic school or department. All of the QUBIS Limited companies began operations either on or adjacent to the University Campus. Seven were initially accommodated in laboratories or space attached to the academic departments with which they were most closely associated. This enabled equipment to be shared between the companies and the academic groups. One unexpected benefit of having access to company-owned equipment was that the University academics were able to win research contracts which they would not otherwise have bid for. As well as having the use of the space and university facilities, the companies had the benefit of easy access to scientific advice and technological expertise.

Four of the companies soon moved out as they expanded and outgrew the space available. One of these, Audio Processing Technology Limited (APT), has now made three growth-related changed of premises. Three companies started up within the QUBIS Limited building and one of these, KAINOS Software Limited has since made two moves to cope with successive stages of rapid expansion, the Company's new premises however remain close to the campus.

QUBIS Limited believes that it is important to provide accommodation for embryonic companies close to their research base and wherever possible to help them find accommodation close to the University as they expand, thus providing many of the benefits of a Science Park without an investment in real-estate. This becomes increasingly difficult as more companies are formed. The need for incubation space has prompted the University seriously to investigate the possibility of becoming involved in a Science Park venture adjacent to its campus.

The QUBIS Group of Companies

Overview

To mid-1995 17 companies had been formed, two had been sold and one had ceased trading. The combined turnover of the current portfolio exceeded £8m in 1995 and virtually all of this was for exports outside Northern Ireland. In 1994 one company won a Queen's Award for Export Achievement and collectively the companies have won 16 DTI SMART awards for innovation. One hundred and seventy staff are employed, most of whom are graduates. None of the companies has failed to date, defying the statistical experience of business start-up and especially of high-technology start-ups (Ganguly, 1984). All except the most recently formed companies are operating profitably.

Textflow Services Ltd

The first company to be formed arose from contract research which had been carried out by the University's Computer Science Department for a local printing company, W & G Baird Ltd. This was to enable the output of any word-processor, to be used to produce camera-ready copy for printing without the need for additional type-setting. Not only was this quicker than traditional methods but, because laborious re-keying was eliminated, so too were the opportunities for errors to occur.

At the time of its development, this facility was not available commercially elsewhere and W & G Baird found themselves with the opportunity to introduce a unique new service. However, it was felt that what was envisaged would not easily sit alongside Baird's traditional printing business. This, coupled with the uncertainty of demand for the new service, suggested that it would be better developed as a separate entity. Another consideration was that of timing. For commercial reasons it was desirable to introduce the new services as state-of-the-art - a 'first' for Northern Ireland, even though it was recognised that further technical developments and enhancement would be needed. University scientists would have an ongoing role in developing the product in the light of operating experience. Textflow inherited from one parent the ability to market the new technology while working with the other to extend that technology so that additional features could be offered.

Textflow Services Ltd was formed in 1985 with equal shareholdings of £28,000 provided by W & G Baird and QUBIS Limited and assistance from the Industrial Development Board for Northern Ireland (IDB). The Company began trading in 1986 employing two staff and reached a financial break-even position in its third year of trading.

Although the original technology has now been overtaken, Textflow Services has continued to grow by building on the early lead established and by adapting and licensing-in the latest technology. Turnover in 1995 was about £1.3m. In 1992 QUBIS Limited sold its shareholding to W & G Baird and the University.

PL Coatings Ltd

Arising out of expertise in the University's Physics department, dating back to 1966, PL coatings was formed to design and manufacture thin film optical coatings for lasers and optical systems. The decision to form the company followed the success of the University's Physics Department in supplying special purpose optical coatings for lasers to meet the very high specifications required by research laboratories. In particular the researchers had developed the expertise to control the spectral characteristics of the coatings to very precise tolerances. Initial customers of the Physics Department included British Telecom and the Rutherford Appleton Laboratory. A market survey commissioned by QUBIS Limited indicated that there would be sufficient demand to produce coatings for laser and optical communication systems and on the basis of this assessment PL Coatings was formed in 1986.

The Company sells its products internationally. QUBIS Limited owned 90 per cent of the equity and the original University researchers, one of whom was the first General Manager of the Company, held a 10 per cent stake. Following the departure of one of the original researchers, additional shareholders became involved and QUBIS Limited eventually sold its remaining stake in 1993.

BIOSYN Ltd

The growth in the use of synthetic peptides and antibodies for use in medical diagnosis and treatment, led to the formation of BIOSYN in 1985, originally 100 per cent owned by QUBIS Limited. The company was formed to produce special purpose peptides and antibodies for research laboratories.

Although initial sales were encouraging much of the potential market soon disappeared - largely as a result of the increasing a availability of low cost peptide synthesisers - as research laboratories acquired the equipment to produce their own sequences. In the meantime however the company had developed good contacts with research laboratories worldwide and was able to reassess the changing market needs and match them with appropriate expertise within the company. As a result, BIOSYN decided to focus on the design and development of diagnostic aids and kits. The considerable potential identified for its diagnostic kits coupled with the unavoidably long gestation time to test and bring such products to the market necessitated additional investment and in 1994 additional share capital, including venture capital, was raised. The current holdings are QUBIS Limited 66 per cent, management and staff 11 per cent and the venture capitalists 23 per cent. In 1996 Biosyn obtained approval from the US regulatory authorities to sell its Gastrin Enzyme Assay kit in the USA and has now entered into an marketing agreement with an international supplier of pharmaceutical products.

MarEnCo Ltd

MarEnCo provides international environmental consultancy services and is 27 per cent owned by QUBIS Limited. Other shareholders include management and staff, 18 per cent and prominent local firm of consulting engineers, 51 per cent. MarEnCo was formed to provide a commercial outlet for a growing number of requests to the University's Marine Biologists to undertake overseas consultancy. Although this type of work could be obtained by the University acting purely as an academic institution there was a perception that it could compete more effectively if professionally marketed and also that more favourable contracts could be negotiated by a commercial company. MarEnCo began with equity shared between QUBIS Limited and several key members of University staff but later a shareholding was offered and taken up by a local firm of consulting engineers. Unlike most of the QUBIS Limited companies, MarEnCo's services are at the 'soft' end of the spectrum.

KAINOS Software Ltd

KAINOS is a Greek word meaning new, fresh or innovative. KAINOS Software Limited, which designs and supplies office automation software, arose from a close relationship which had developed between the University and one of its computer suppliers, ICL. The availability of a ready pool of graduates from Queen's coincided with a need by ICL for software design and development.

QUBIS Limited holds 49 per cent of the equity, ICL took a 51 per cent stake in KAINOS and the Company grew rapidly, firstly producing products for ICL/Fujitsu and subsequently for other customers as well. The original technical focus was on Unix based systems and programming in the 'C' language with a bulk of the activity leaning towards systems rather than applications software. The Company places a high priority on work methodologies and documentation standards in the knowledge that almost all of its software will be exported and sold to eventual customers by agents. As the Company has grown the range as well as the quantity of work has grown. There are close links between KAINOS and the University's Department of Computer Science. Annual sales turnover is approaching £3m, more than 95 per cent of which is exported. Total employment is 111 and KAINOS has purchased its own premises very close to the University

QUBECON Ltd

QUBECON began as a specialised economics consultancy. The intention was to sell University expertise, mainly to public sector clients, in competition with major consultancies. The main areas of expertise were in economic forecasting and modelling for industry, commerce and the public sector; market research of retail markets and price structures and international price and income comparisons; environmental impact studies in accordance with EU directives for projects related, for example, to energy, tourism, transportation and water resources.

Shareholdings were QUBIS Limited 50 per cent and 25 per cent each by two founding members of academic staff. QUBECON relied heavily on the personal commitment and expertise of these two University researchers. When one of these left the University after less than two years, the Company ceased trading.

DYNOS Turbines Ltd

DYNOS is developing low cost gas turbine engines for general use. The Company was formed to exploit research being carried out for the aerospace industry by mechanical engineers at Queen's. A low-cost turbine engine had been developed for use in drone (target) aircraft and the potential for its more general use was apparent, e.g. as a replacement for diesel engines in standby generators and for use by the bulk powder transport industry as an on-board means of loading and unloading powders to vehicles. The main shareholders are QUBIS Limited 15 per cent, University researchers 42 per cent and a private company 40 per cent.

Audio Processing Technology (APT) Ltd

APT was formed to exploit University-held patents related to the electronic compression of digitally encoded music signals. These techniques enable music to be coded without loss of fidelity using four digital bits instead of sixteen. QUBIS Limited managed to attract the interest of a major international company, a subsidiary of Carlton Communications, to take a 51 per cent stake. The University researchers also have shares in the company, owning 29 per cent and QUBIS Limited retains 20 per cent.

By producing chips for OEM equipment manufacturers, APT has exported most of its products and become widely known through the use of its chips in cinema sound equipment for such prestigious productions as

'Jurassic Park' and 'Schindler's List'. The Company which is now the fastest growing of the QUBIS Limited companies (excluding embryonic firms) whose sales in 1995 exceeded £3.3m has now moved to prestigious new premises in Belfast.

Lumichem Ltd

This company is developing and exploiting a revolutionary amino-acid detector reagent. This patented reagent, whose manufacture is closely controlled by Lumichem, fluoresces at a specific wavelength of light. The reagent and light sources are sold via agents to research and forensic laboratories throughout the world. Lumichem's unique reagent was heralded as a breakthrough by its specialist users and the Company was greatly encouraged by the rapid take-up of its product including a number of pre-paid orders for early supplies. There has been a relatively low capital investment in Lumichem which has shown a healthy return. Shareholdings are divided equally between the management, QUBIS Limited and a private UK investor.

ANDOR Technology Ltd

Work in the University Physics Department led to a requirement, for research purposes, of pulsed light spectroscopy devices. Although the researchers spotted the potential for wider application of these units, it took some time to interest a commercial company in their development. After several false starts a major US instrument manufacturer, the Oriel Corporation, took an interest and a 51 per cent stake in ANDOR. Considerable development was required to develop a laboratory instrument into a commercially saleable product but now ANDOR's Diode Array Multichannel Analysers are in use worldwide. Growth has since been rapid, 26 staff were employed in 1995, sales are approaching £2m and the Company has moved to new premises from the university department where it began operations in 1989. Management and staff hold 42 per cent of the Company shares and QUBIS Limited retains 7 per cent.

Integrated Silicon Systems (ISS) Ltd

Taking advantage of the trend towards more complete integration of electronic circuitry and building on expertise in the University's Electronic Engineering and Semiconductor laboratories, ISS designs Application

Specific Integrated Circuits (ASICs). Specifically it manufactures Digital Signal Processing (DSP) ASICs for use in telecoms, broadcast, computer, aerospace, consumer and medical applications. Uniquely ISS offers a process based on a design concept of 'algorithm-to-architecture-to-chip' design flow. This mirrors the silicon methods used in development of DSP software algorithms and allows designers rapidly to translate their DSP algorithms into dedicated silicon chips. An example of the Company's capability is its design of a video processing chip which can process over 5 billion multiplications and additions per second (5,000 MIPS), measures less than 1cm by 1cm and has a power dissipation of less than 1 watt.

The Company has successfully carried out development work for clients in the UK and North America. The DSP chip market is estimating to have an annual growth rate of around 30 per cent with increased demand in telecoms, multi-media disc-drive and broadcasting. The Company is 60 per cent owned by management and University staff, 20 per cent is held by venture capitalists and 20 per cent by QUBIS Limited.

Rotosystems Ltd

During research into the variables which affect the process of plastic rotational moulding, which is used in the manufacture large hollow objects such as plastic oil tanks, it became necessary to make temperature measurements at critical stages inside the rotating moulding oven. A rugged sensor and telemetry signalling system, which could withstand the harsh conditions inside the oven, was designed and built for research use. The results of this work gave hitherto unknown information about critical stages of the moulding process which could be used to provide very precise control over what had previously been very much a 'hit-or-miss' process. This led to greatly improved quality control of the finished products and associated economies of manufacture.

The commercial implications were appreciated by the researchers and so Rotosystems was formed to manufacture and market a production version of the laboratory instrument, known as the 'Rotolog', to users of rotational moulding equipment worldwide. Rotosystems is owned 34 per cent by QUBIS Limited, 44 per cent by management and University staff and 22 per cent by a local plastics development company.

Hughes and McLeod Ltd

Hughes and McLeod is one of the few companies in the group arising from an idea which originated outside the University. The concept of plastic reeds to replace natural material, traditionally used for bagpipe chanters, was developed by the original entrepreneurs Messers Hughes and McLeod. They came to the University's plastics experts, originally for assistance with testing and later to improve the design of their reeds, which are now increasingly being used by Highland pipers. A member of University staff now has a 5 per cent stake, QUBIS Limited holds 16 per cent and 79 per cent is held by the management and a private investor.

Vinifer Ltd

Vinifer, which is owned 51 per cent by QUBIS Limited and 49 per cent by its management, produces chemical monitoring systems for testing laboratories. The need was recognised for quantitative measurement of markers in rebated hydrocarbon fuels, e.g. red diesel fuel. These tests are required as a matter of routine, for example, by customs offices and oil refineries. Currently available tests were time consuming and inconvenient involving the handling of hazardous chemicals by personnel, often under adverse weather conditions. Furthermore the procedural use of large volumes of test solutions adds to the complexity of these tests which limits their success. Vinifer's solution to this problem has been to develop and supply a stand-alone instrument which performs quick, accurate and low-cost quantitative analysis of commonly used markers in hydrocarbon fuels. In one of the first of a growing number of inter-company collaborations within the QUBIS Limited group, the initial product was the result of collaboration between Vinifer's current Managing Director who is a chemical engineer, another QUBIS Limited company, Andor Limited, and the University's School of Chemistry.

Biocolor Ltd

Biocolor produces assay kits, mainly for collagen and elastin, for use in tissue culture laboratories. The original idea came from the personal experience of a researcher in the University Biochemistry Department who owns 40 per cent of the equity. QUBIS Limited holds 60 per cent. This is one of the few QUBIS Limited companies which does not have a commercial partner and, possibly as a result, growth has been relatively slow and sales to date have been modest. Most kits sold so far have been to Japan but they have aroused a great deal of interest amongst international pharmaceutical companies and many leading research laboratories. There is believed to be considerable potential for growth and a new marketing campaign aimed at the United States is just getting underway.

Osprey Environmental Ltd

The most recently formed company in the Group, Osprey is developing personal electrical and magnetic field monitoring equipment for both professional and consumer use. The 'Live Alarm' is intended to provide a warning to personnel working with or near live electrical equipment. It can be clipped to a coat or belt and gives an audible alarm as the wearer approaches hazardous equipment. This is an example of an idea which arose from a 'throw-away' remark by an electrical engineering lecturer which was picked up by an alert PhD student who could see its commercial potential and who subsequently had the enthusiasm and energy to champion the idea. QUBIS limited assisted by providing financial advice and a room in its premises where the prototypes were build. The product has already attracted a high degree of interest and some early orders have been received. The management i.e. the one-time PhD student - holds 75 per cent of the equity and QUBIS Limited the remaining 25 per cent.

Private Sector Partners

The private sector investments in the companies is shown in Table 12.1.

Table 12.1 Private sector investment in QUBIS Limited companies

Company	Private sector shareholdings
KAINOS	51% by ICL Computers plc
DYNOS	40% by private company
APT	51% by a Carlton Communications company, Solid State Logic Ltd
MarEnCo	51% by Kirk McClure and Morton, a prominent local firm of consulting engineers
ANDOR	51% by Oriel Corporation, USA
Lumichem	33% by a private investor
Rotosystems	22% by Plastics Development Centre, a NI based company
ISS	20% by a private investor, 20% by Innovation Equity Ltd (a venture capital fund)
Hughes & McLeod	74% by private investors
Vinifer	49% by a private investor
Osprey Environmental	75% by a private investor
Biosyn	23% by Enterprise Equity (NI) Ltd (a venture capital fund)

Age Distribution of Companies

The holding company and five others have been in operation for ten years or more. Seven companies have been operating for between five and ten years; two for four years and the most recently formed is in its second year of operation (Table 12.2).

Table 12.2   Age profile of QUBIS Limited companies

Company	Founded	Age (1996)
QUBIS Ltd	1984	12
Textflow	1985	11 (sold 1992)
PL Coatings	1985	11 (sold 1993)
BIOSYN	1986	10
KAINOS	1986	10
Marenco	1986	10
APT	1988	8
QUBECON	1988	2 (ceased trading 1990)
Dynos	1989	7
Lumichem	1989	7
ANDOR	1989	7
ISS	1990	6
Rotosystems	1990	6
Hughes Macleod	1990	6
Vinifer	1992	4
Biocolor	1992	4
Osprey	1994	2

'Tripod' Profile of Companies

Table 12.3 shows the Tripod profile scores of the companies for the period immediately following start-up and currently.

Extent of University Control and Involvement

All the companies score either four or five on this parameter, the differences arising from those cases where the University, through QUBIS Limited, is a majority shareholder and those in which it has a smaller stake. In all of these spin-outs Queen's has been proactive in seeking opportunities to exploit its research. None has occurred fortuitously or without the vigorous involvement of the holding company.

Table 12.3 Tripod profile for QUBIS Limited companies at formation and currently

Company	Extent of University control		University staff involvement		Products or services offered	
	When Formed	Now	When Formed	Now	When Formed	Now
QUBIS Limited	5	5	5	4	5	5
Textflow	5	4	5	4	5	5
PL Coatings	5	0	5	0	5	5
BIOSYN	5	5	4	4	5	5
KAINOS	4	4	4	4	5	5
Marenco	5	4	5	4	3	3
APT	4	4	4	4	5	5
Dynos	4	4	4	4	5	5
Lumichem	5	4	3	3	5	5
ANDOR	4	4	4	4	5	5
ISS	4	4	4	5	4	4
Rotosystems	4	4	4	4	5	5
Hughes Macleod	4	4	4	4	5	5
Vinifer	5	5	3	3	5	5
Biocolor	5	5	5	5	5	5
Osprey	4	4	2	2	5	5

Nature and Extent of University Staff Involvement

Most companies score four, indicating the presence of university staff as shareholders and/or directors in the companies. There are presently two companies which have University staff members in executive roles, one of these is on secondment and the other is a part-time activity. At start-up the then General Manager of QUBIS Limited was a member of University staff and at least three other companies had university staff in either full-time or part-time roles at early stages in their development but as the companies grew these people were all replaced with full-time company employees. Two companies score three at present indicating that they receive paid assistance from University staff, generally on a consultancy basis. One fledgling company still receives unpaid assistance and scores two. All of the companies have from time to time made use of university staff as consultants for which they have paid commercial rates, usually negotiated by the University's Industrial Liaison Office.

Products or Services Offered

Virtually all the companies are producing 'hard' products or services which require a commitment to significant expenditure of both direct and indirect costs. Almost all are also export orientated. There is one consultancy company which scores three and one other, scoring four, which does not have a product-line as such but which offers specialised services.

Overall, the profile of this group of companies is that shown in Figure 12.1. The companies operate predominantly as providers of products and services which are at the outer end of the spectrum; there is considerable institutional involvement and control on the part of the University (via the holding company) and the participation of University staff is mainly as shareholders and directors.

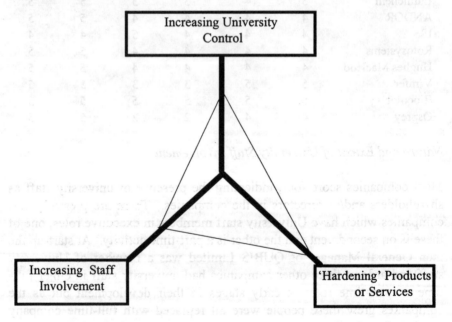

Figure 12.1 Average profile of QUBIS Limited companies

It is emphasised that the purpose of the model is to indicate a profile which will be indicative of the type of system which had to be put in place to help establish the companies in the first place and then to sustain them and help them to flourish.

Company Growth

Figure 12.2 illustrates the combined growth in turnover, in numbers employed and in Profit Before Tax (PBT) for an eleven year term (including those for Textflow Services). The 1996 turnover figures are estimates. There are strong correlations between turnover and numbers employed (r = 0.99) and between turnover and PBT (r = 0.94).

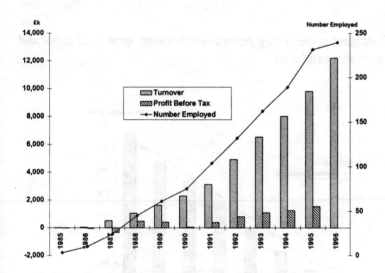

Figure 12.2 Combined turnover, profit before tax & numbers employed

The five longest established companies have shown continuous growth of turnover and numbers employed, see Table 12.4 and Figure 12.3. The Table shows rapid initial growth which, although tailing off, is still continuing after ten years. One of these companies, in biotechnology, has had a sustained period of investment in product development which is not unusual in that industry. It began a period of growth in its fifth year. Four companies are at very early stages and one is virtually dormant.

Table 12.4 Annual growth of the five longest established QUBIS companies

Year of operation	Turnover	Employees
2nd year	125%	81%
3rd year	75%	39%
4th year	34%	17%
5th year	30%	32%
6th year	27%	22%
7th year	26%	19%
8th year	16%	13%
9th year	14%	15%
10th year	15%	7%

Figure 12.3 illustrates the strong correlation between growth of sales and increase in numbers employed.

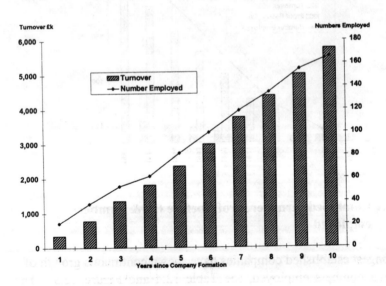

Figure 12.3 Combined turnover and employment in the five oldest companies

Because the companies were formed at various times throughout the past ten years, it is quite difficult to extrapolate the growth rates of the five oldest to the entire group. Figure 12.4, however, shows annual percentage growth of combined turnover and employment by the years since each company was formed. Thus there are 16 firms which have been in operation for one year and progressively fewer for longer periods. This data needs to be interpreted with caution since the calendar years are different - for example one of the companies reported in year 1-2 was formed in 1985 and another in 1995. Nevertheless it does show a decreasing rate of growth as companies age. This growth pattern is consistent with the results of studies, referred to earlier, by American Electronics Association AMA (ibid.) and Morse (ibid.), that young innovative firms grow faster than mature firms and that they have a high potential to create jobs. The survival rate for this group is also bucking the trend for new technology-based start-ups and is certainly better than the 30 per cent mortality rate after three years observed by Ganguly (ibid.).

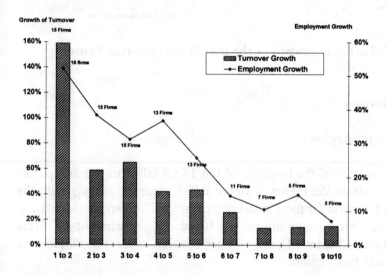

Figure 12.4 Year-to-year growth of turnover and numbers employed

The growth pattern of the companies varies. Figure 12.5 shows the rapid growth in turnover of the four fastest growing firms. Five others have grown steadily but more slowly.

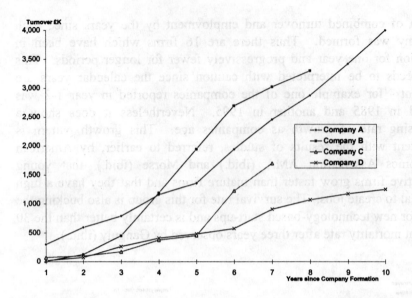

Figure 12.5 Turnover growth of the four fastest growing firms

Economic Impact

Exports from the Region

Almost 90 per cent of the turnover of the 14 QUBIS Limited companies comes from outside the region, a figure which contrasts sharply with the proportion of exports for most small companies in Northern Ireland. Even amongst the local non-university based 'high-technology' firms interviewed (see Chapter 14), only 41 per cent had more than 80 per cent of sales outside their region.

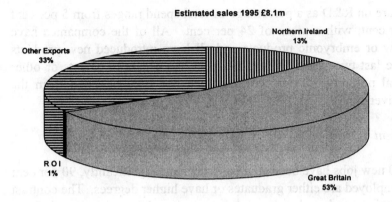

Estimated sales 1995 £8.1m

Other Exports 33%

Northern Ireland 13%

ROI 1%

Great Britain 53%

Figure 12.6 Destination of sales for 14 QUBIS Limited companies in 1995

KAINOS Software is distributed throughout the world by ICL. Lumichem has sold its amino-acid detector in 20 countries. ANDOR has sold to USA, France and Germany. BIOSYN has sold diagnostic products to large US drug companies. APT's music coding system is being used all around the world as its chips are incorporated in satellite transmission systems and long range radio links. The polycarbonate bagpipe reeds produced by Hughes and McLeod are used in USA, Canada, Oman and Australia as well as in Scotland. Rotosystems' product Rotolog is used in Germany for monitoring the quality of motor cycle fuel tanks as well as throughout Europe and USA. Biocolor sells more of its Sircol assay kits to Japan than to anywhere else. The outlook of the companies is truly international so it is not surprising that more than half of the partners in QUBIS Limited companies are based outside Northern Ireland (Cartin, 1995).

Expenditure Patterns

About 80 per cent of expenditure is within Northern Ireland, including 98 per cent of salaries. The economic impact of a local expenditure of some £5.5m is bound to be increased due to the multiplier effect which has been estimated variously, for high-technology firms, at between three and five.

Commitment to Innovation

Expenditure on R&D as a percentage of total spend ranges from 5 per cent to 80 per cent, with a mean of 24 per cent. All of the companies have either new or embryonic products and all have introduced new products during the last two years. Ninety per cent had access to patents or other intellectual property, 80 per cent of which had been obtained from the parent university.

Employment

Some 180 new jobs have been created to date and significantly, 90 per cent of staff employed are either graduates or have higher degrees. The contrast between university-based and non-university-based firms has already been noted and the contrast between the QUBIS Limited companies and the other university-based firms in the sample interviewed (see Chapter 13) is also marked, some 90 per cent of staff in the QUBIS Limited companies being either graduates or holding higher degrees compared with 72 per cent for the university company sample overall.

Key Tasks (Sub-Systems)

Given the profile of the relationship between the QUBIS Limited companies and the University, which tends towards the outer extremes of each parameter, see Figure 12.1, there are a number of key tasks which are particularly important. The approach taken to these by QUBIS Limited has evolved with experience.

Identifying Prospective Projects

There is no regular trawl for exploitable research, nor are researchers required to make returns of their research to QUBIS Limited. There is however a fairly effective networking system based on personal contacts. Some projects are bought directly to QUBIS Limited by researchers who are keen to form companies to exploit their ideas and inventions, some are referred by the University Industrial Liaison Office as a possible alternative to other means of exploitation such as licensing or collaborative research. It is arguable that the lack of a more formal system for identifying prospects may result in potentially valuable opportunities being lost.

Assessing Commercial Potential

Shortly after the formation of QUBIS Limited there was an upsurge of interest and a number of potential projects were offered for evaluation. These were mainly of the technology-push variety from university researchers but, as is usual in this type of situation, only a small proportion of the projects proposed survived detailed technical, economic and market analyses. Nevertheless, several demand-led ideas were also identified by local companies.

There is seldom a need to second guess the technical feasibility of products proposed by University researchers but, for the first few companies formed, quite formal market research was commissioned and although this was useful, in that it provided an independent view, it was soon abandoned in favour of a more pragmatic approach. The procedure now is to try and interest a potential and experienced business partner, ideally a company with a market presence and distribution network. Experience has shown that persuading such a partner to take a stake is sufficient evidence that a market either exists or can be created for the idea. Experienced industrial partners are now equity holders in the most successful QUBIS Limited companies. It is worth noting that to attract industrial partners to a joint venture it is often necessary to offer them a majority shareholding. QUBIS Limited is quite sanguine about this as it ensures that the industrial partner is committed to take first responsibility for the new project.

Deciding Whether and When to Form Companies

Even when commercial viability is apparent the decision to form a new company is not automatic. Apart from the considerations which would be addressed by any potential entrepreneur, it is at this stage that the influence of the wider University system has an impact. Alternatives are considered, such as licensing-out the technology or engaging in collaborative development with an existing firm. In some cases, for example, the desire of researchers to publish may be at odds with the need to obtain patents which are sometimes necessary if the research is to be commercialised. The value or prestige to a University School of industrially funded research may indicate a preference for a collaborative research contract.

Putting in Place Entrepreneurs, Product Champions and Full-time Managers

Although, in hindsight, it is possible to identify key individuals who could be regarded as entrepreneurs, this is by no means always clear when a project is being evaluated. Frequently the project is proposed by a product champion or 'missionary' who is determined that it should succeed although these people often lack the necessary business skills and commercial acumen. In a number of the QUBIS Limited spin-outs, entrepreneurs have been recruited to manage and drive projects forward and in others the holding company has provided support, either directly by way of practical assistance or indirectly by nominating experienced Board members whose skills complement those of the academics involved.

It soon became apparent that to obtain the necessary commitment it was necessary for all of the companies to have full-time managers. Academic researchers have in some cases been retained as advisers and consultants but, despite some earlier pressure to do so, no academics have been appointed to managerial roles.

Reconciling Objectives and Expectations

This problem was recognised at an early stage, in at least one spin-out firm, when it became apparent that the main objective and expectation of one academic board member was to obtain equipment for his laboratory whereas that of the University was to obtain a foothold in the marketplace. In the short term these aims were incompatible and as a result the professor resigned from the company which he had helped to form. QUBIS Limited learned from this experience and now there is a very thorough exploration of expectations and motivations before a new company is formed.

Interaction with Academic Departments

Relationships between spin-out companies and academic departments are, by and large, good but there is potential for discord. Again, one or two early experiences have proved salutary. Partly this was due to commercial naïveté on the part of some university staff and partly because of unrealistic expectations by the company staff about academic commitment to the new firms. Considerable efforts are now made, for example, to ensure that where university resources are used they are paid for, or alternatively that there are quid pro quo arrangements for academic staff to use company equipment and facilities. It is also regarded as being very important to give formal and 'public' recognition of support given by the University departments. Some of the best examples of collaboration have occurred because the companies have helped the academic departments win research funds and credibility with sponsors which they would not otherwise have obtained.

Interacting with the Wider University System

This has been achieved through participation by senior University staff on the Board of Directors. The Vice-Chancellor is currently Chairman of the Board of the holding company which also includes the Bursar, and three senior members of academic staff. There is also close collaboration between the staff of the holding company and those of the University's Industrial Liaison Office with which it shares office accommodation.

Establishing Boards of Directors

QUBIS Limited is represented on the Boards of all of the spin-out companies, usually by a member or past member of the holding company and in six of the firms QUBIS Limited has supplied the board Chairman. An effort is made to provide a balance of skills and expertise. Usually the technical input comes from the researchers or product champion but sometimes it is necessary to temper their natural enthusiasm through the presence of other experts who can ask searching questions - and understand the answers given. Usually it has been necessary to introduce financial expertise, mainly through the holding company in the first instance. Commercial judgement and marketing know-how is most appropriately provided by the industrial partner to the venture. This indeed is one of the greatest strengths of the corporate venturing approach, industrial partners bring much more than simply funds to the transaction. Many have established distribution systems through which the products of the new firms can be sold.

Recognising and Dealing with Potential Conflicts of Interest

Conflicts of interest have only rarely occurred in the QUBIS Limited firms. In one or two early cases there were apparent clashes between the interests of academic departments and companies in which University staff were working part-time. One debate hinged on whether a company was undertaking work which would otherwise have been done by an academic school. In that case the head of the academic unit was invited to attend company Board meetings in an effort to reassure him that there were mutual advantages in collaborating. In several cases it was necessary to agree that payment for University resources would be made by companies and would be negotiated independently by the Industrial Liaison Office.

In this context it is worth noting that there was a learning experience not only for academics involved but for the holding company staff who were not ex-academics. At one stage there was a culture clash when logical arguments were being employed to no effect by company staff to counter emotional issues being raised by academic staff.

Accessing Finance

This task has fallen mainly to the holding company in consultation with government agencies and industrial partners. QUBIS Limited makes modest financial investments of between £5,000 and £50,000 in new businesses. Northern Ireland has a favourable grants regime to encourage new company formation and growth. In the early stages of the venture QUBIS Limited received £ for £ funding to help assess new projects and this helped in the formation of the first spin-out companies. Since adopting the corporate venturing approach, funding from government sources has been matched by contributions from industrial partners. It can sometimes be difficult to obtain external 'seedcorn' funding of a few thousands of pounds to demonstrate the viability of projects. Once the initial viability of a project or product has been established however, funding for new company formation has not proved to be a limiting factor. The DTI SMART award scheme has proved particularly valuable in helping take products through their initial stages. The QUBIS Limited companies have won a total of 16 of these competitive awards since 1988.

Development capital has come in most cases from additional investment by the shareholders and from retained profits. Two firms have entered into agreements with venture capital funds in exchange for 23 per cent and 20 per cent of their equity.

Assessing Performance

Unlike many of the firms surveyed elsewhere, there is no ambiguity whatever about the performance yardsticks for the QUBIS Limited companies. To quote the Chief Executive, Edward Cartin

> Ultimately, the company wants to make a profit, ... we are looking to investments to make profits such that they can pay us the dividend. That is the prime consideration. There is the temptation to judge success on employment and other criteria but that is a by-product of profitability (Cartin, 1994).

To what extent has this aim been met and what have been the benefits to Queen's University? The University's total cash investment in QUBIS Limited was less than £500k. Apart from any other benefits which may accrue, the valuation of the investments now held by QUBIS Limited, under the most conservative accounting basis, exceed £1m. On a more realistic basis, such as that used by venture capital companies in valuing their investment portfolios, these holdings would be assessed at between £2m and £4m. Assuming a valuation of say £2m this represents a very respectable average compound rate of return on capital over nine years of about 17 per cent.[1]

Taking as an example KAINOS Software Limited in which QUBIS Limited has a 49 per cent stake; the QUBIS Limited investment in that Company, in 1986, was £25k. The KAINOS balance sheet at 31 December 1995 showed the total shareholder's funds to be £1.8m. At this very conservative valuation the QUBIS Limited share is worth just under £900k. An alternative way of viewing this is to note that KAINOS Limited's profits are currently running at about £500k per year before tax. At a modest price to earnings (P/E) ratio of say six - which is unusually low for a technology-based firm - the company would be valued at £3m making the QUBIS Limited share about £1.5m.

Although dividends have been paid to QUBIS Limited from a number of its companies none have yet been paid by QUBIS Limited to the University, emphasising that growing companies are cash-hungry and underlining the long-term nature of this approach to managed exploitation. Even though there has not yet been a dividend stream, there is little doubt that, even in purely financial terms, this has been a good investment for the University. There have of course been other inputs including the time of University staff and others, not least the unpaid external members of the

QUBIS Limited Board many of whom also serve on the Boards of the various companies. Nevertheless the opportunity costs, represented by alternative investments foregone, would be most unlikely to offset the benefits of the appreciating asset represented by the current portfolio of shareholdings. There are, additionally, less tangible benefits to the University in terms of its prestige and public visibility and being seen to contribute to the local economic community.

Rewarding Inventors/ Researchers

Almost always the rewards for inventors have been through the acquisition of shares and directorships, often in exchange for quite nominal sums invested. Every company in the group has at least one member or former member of University staff as a shareholder. There have been no cash millionaires to date although one or two have, on paper at least, accumulated considerable wealth. In a few cases researchers are also employed as paid consultants to the companies.

Supporting Embryonic Companies

QUBIS Limited is a rather unusual type of holding company - not only does it create new companies but it continues to provide support and practical assistance for as long as is required to its offspring. Apart from initial help at set-up the aftercare extends to activities such as central provision of accounts including VAT and payroll. Help is also available in the early stages with marketing and advice is always available on matters such as budgeting and forward planning. Quite soon most companies outgrow the need for this level of intensive care and recruit their own specialist staff. QUBIS Limited makes only nominal charges to embryonic firms for these services which not only greatly increases the confidence of novice entrepreneurs but also their credibility when they deal with creditors, banks and customers.

Discussion and Conclusions

The record of new company formation and growth is impressive, as is the survival rate of these firms. They have undoubtedly contributed significantly to the meagre agglomeration of high-technology firms in Northern Ireland, and have taken more DTI SMART awards than any other grouping. Their growth, as expected, conforms to the variation of the Pareto principle with about 80 per cent of the employment and turnover being accounted for by three of the 17 companies. Furthermore the corporate venturing methodology pioneered by QUBIS Limited is being emulated by a number of UK universities.

Although the closely managed hands-on approach has produced durable companies the question must be asked whether there is a downside to this model. Queen's, in common with most UK universities, adopts a possessive attitude to the ownership and exploitation rights of its intellectual property. It is possible that this proprietorial approach together with the need for them to operate via QUBIS Limited deters some potential entrepreneurs. The number of companies emanating from say the University of Linköping for example, which has a much more relaxed attitude to spin-outs, greatly exceeds that from Queen's.

It is becoming increasingly difficult for QUBIS Limited to find new opportunities which meet its rigid standards for investment. Furthermore the criteria for investment are becoming progressively tighter and some smaller enterprises which would have been funded in the earlier years would no longer pass muster. This is not an argument for lowering standards but to question whether there is room for a trade-off between the number of projects accepted and the subsequent failure rate. To have had no companies fail in more in a decade could suggest an overly cautious approach. Although the number of 'unauthorised' companies is not thought to be large, some instances are known of entrepreneurs deliberately avoiding the QUBIS Limited system and forming companies by other means.

As a matter of practical policy QUBIS Limited does not encourage students to come forward with projects, compared with say University of Twente which has an impressive record of student-based spin-outs. The QUBIS Limited approach may make sense from the point of view of the University but from the viewpoint of helping develop the local economy it discriminates against a large constituency of potential entrepreneurs.

From a regional economic perspective one might also ask whether more could be done to assist those whose ideas are not taken up by QUBIS Limited. Inevitably the proportion of ideas taken up by QUBIS Limited is fairly small. Although there are plenty of technological opportunities only those which meet the criteria of being able to demonstrate that there is either an existing or a potential market niche are likely to be accepted. A further requirement is for a 'missionary' with the ability to perceive a market need and the energy to fulfil it. The Chief Executive of QUBIS

Limited is a believer in Peter Drucker's maxim that - 'Anything worthwhile in this world is usually done by a monomaniac with a mission'.

Another characteristic is the relatively low level of investment made by QUBIS Limited in its offspring. This is due partly to the ability to match funds from industrial partners with government funding but it also may be indicative of a rather timid approach. Contrast for example the investment of a less than £100k in a QUBIS Limited biotechnology-based company with the £7.5m raised on the stock market for a similar company from UMIST. Eventually of course the QUBIS Limited investment may prove to have been the more prudent but there is something in the argument that one must take bigger risks to obtain higher rewards.

Summary

QUBIS Limited was established in 1984 charged with identifying and exploiting, commercially, University research through new company formation. The Company is wholly owned by the University, has an authorised share capital of £500,000 of which £381,000 has been issued. The immediate objective of QUBIS Limited is to produce a stream of dividend or other income from its investments.

Seventeen technology-based companies have been formed, two have been sold and one has ceased trading. No company has failed to date and all are operating profitably. Growth has been continuous, 170 staff, mostly graduates and postgraduates are employed in the fourteen companies which remain within the group. The combined sales turnover of the group exceeds £8m. Almost 90 per cent of the turnover is exported and about 80 per cent of expenditure is within Northern Ireland, statistics which indicate the extent of the contribution to the regional economy.

Almost all the companies score five on the Products arm of the Tripod, indicating that they are product-based as distinct from soft companies. The QUBIS Limited holding varies from 15 per cent to 90 per cent and most companies score four or five on the University Control/Involvement arm of the Tripod. Most companies score four on the Extent of University Involvement arm, indicating the presence of university staff as shareholders and/or directors.

A particular characteristic of the QUBIS Limited approach is its preference for a form of Corporate Venturing through which it prefers to enter into equity-sharing joint ventures with existing, established companies. This provides not only funding but expertise, access to markets and complementary expertise.

A possible downside to the QUBIS Limited approach is that its rather prescriptive approach may possibly deter some potential entrepreneurs, although it is recognised that a more liberal approach could affect the very low failure-rate which is currently achieved. From a regional employment perspective one might also question the policy of not inviting projects from students and one could also ask whether more could be done by either QUBIS Limited or the University to help exploit the technological ideas which do not meet the criteria to be taken up as QUBIS Limited projects.

Note

1. In late 1996 QUBIS Limited sold its shareholding in APT Limited for about £1m. The QUBIS Limited investment, in 1988/89, was £60k.

13 Performance of University Companies

Characteristics of NTBFs

Since university spin-out companies, consist of mainly new technology-based firms they can be regarded as a subset of NTBFs, the importance and characteristics of which were discussed in Chapter 3. Particularly noted were:

- Their disproportionate share of innovations (Oakey et al., 1988; Rothwell and Dodgson, 1987);
- The association between innovativeness and job creation (Piatier, 1981; De Melto et al., 1980; NBST, 1980);
- Their high expenditure on R&D (Mustar, 1995);
- Their propensity to grow more quickly than low-technology firms (Oakey and Rothwell, 1986; Morse, 1976; American Electronics Association, 1978; Rothwell and Zegfeld, 1981);
- The high mortality rate of NTBFs (Ganguly, 1985);
- The tendency for rapid growth to be confined to a small proportion of any group of NTBFs (Storey et al., 1987);
- Problems related to growth and shortage of key personnel (Rothwell and Dodgson, op. cit.);
- The importance of external inputs (Oakey et al., op. cit.; Lowe and Rothwell, 1987; Roberts and Fusfeld, 1981);
- Their importance in regional development (Vaessen and Wever, 1993; (Keeble, 1993; Kleinknecht and Poot, 1992; Segal Quince Wicksteed, 1986).

Difficulties in obtaining external capital and a potential weakness in marketing were noted in a Swedish study, by Klofsten et al. (ibid.), in which the internal resources of two types of university-based spin-outs,

219

research-based and others, were compared with a non-university group in terms of the resources perceived as being needed at start-up and later.

The Swedish researchers noted that

> A common feature of the groups under study is that they found capital to be in short supply at start and later. Especially the research-based group university spin-offs have experienced a gap between the (perceived) importance and availability of external capital. As to technological know-how, all three groups seem to be rather well off, while there are notable differences for marketing knowledge and personnel. There is a definite gap between the importance and availability of marketing knowledge, both internally and externally , at start for the university groups, while there is a more balanced situation for the non-university group. Firms in the latter group experience a gap between the importance and availability of personnel, which is a lesser problem for the university groups. ... As a total, the non-university group seems to be more 'complete' at start and also adds resources internally as they are needed, thereby having a more balanced situation in their development (Klofsten, Lindell, Olofsson and Wahlbin, 1988).

The companies in the IRDAC study also found difficulty obtaining funds for growth and for vertical integration, (Rothwell and Dodgson, ibid.).

These findings, combined with additional insights arising from the interviews with university practitioners and the experiences of the QUBIS Limited companies described in the previous Chapter, provide the basis for a number of hypotheses to be made.

Hypotheses

It is hypothesised that university-based firms will share many of the characteristics of NTBFs in general - specifically:

- They are likely to be innovative and have a larger proportion of new or embryonic products than low-technology-based firms;
- Compared to more traditional or 'low-technology' companies they are likely to view themselves as competing internationally rather than locally;

- Like other technology-based firms, spin-out companies located in weak economic areas are likely to find their markets outside of their region. They are likely to be export orientated;
- They can be expected to employ a high proportion of graduates in engineering, science and technology and spend a considerable proportion of their resources on R&D;
- They are likely to exhibit above average growth although most of this will occur in a small proportion of firms. There is also likely to be a fairly high mortality rate;
- They are likely to have a realistic impression of how they rate technically vis-à-vis their competitors and other companies in their field.

Because of their links with universities, spin-outs are likely to have some particular advantages over non-university linked companies, namely:

- They will have convenient access to latest technology and external (i.e. non-company) advice;
- They have a ready source of potential graduate and postgraduate employees;
- They are likely to be pre-disposed towards R&D and technologically-based innovation;
- They may have exclusive or privileged access to university intellectual property in the form or patents or unique technological know-how. As a result they may have more technological 'firsts' than non-university companies;
- Although university companies can be expected to exhibit many of the characteristics of the firms in the IRDAC study by Rothwell and Dodgson (ibid.), they are unlikely to be constrained in their growth by shortages of key technical personnel.

University companies may also suffer disadvantages vis-à-vis non-university companies, for example:

- Because of their dependence on the university, they are likely to be less 'robust' for some time after start up and may take longer than other firms to become self-sufficient;

- Evidence from the interviews with university practitioners suggests that university companies may be weaker than non-university-based firms in marketing and may take longer to achieve a significant market presence without external assistance.

To test these hypotheses a second series of interviews was held with university-linked firms, their managers and academic entrepreneurs.

A Second Series of Interviews

The interviews described in Chapter 8 concentrated on the formation, organisation and operation of spin-out companies from the universities' points of view. Although the university staff questioned were able to talk in general terms about the 200 plus companies with which they had been involved, few were able to discuss particular firms in much detail. To test the hypotheses presented above it was necessary to discuss individual companies with those closest to them. A second interview programme was therefore conducted with chief executives and/or academic entrepreneurs of 31 university-linked firms in the UK, the Republic of Ireland and Sweden. The data from this survey was later compared with that from a similar survey of non-university linked firms, described in Chapter 14. That comparison, which is described in Chapter 15, contains most of the analysis and discussion about the similarities and differences between the university and non-university firms. Much of the present Chapter is confined to reporting, rather than interpretation of the data.

The Sample of University Companies

This is not a random sample of university companies, its selection was influenced by:

- Unique access to the companies formed by The Queen's University Belfast (QUB);

- Other UK universities known to be systematically attempting to exploit their technology through new company formation, notably UMIST, Leeds, Manchester, University College London (UCL), and Aberdeen;
- The desirability to include companies from a university in the Republic of Ireland - because of its proximity to and common features with Northern Ireland. Trinity College Dublin was included because of the similarity between its campus companies and the UK based holding companies;
- The fortuitous opportunity to visit companies formed by the university of Linköping in Sweden which has a growing reputation as a seed-bed for spin-out firms;[1]
- The willingness of entrepreneurs and managers in new companies to take part in the study. Some who were approached directly were reluctant to give their time for an extended interview and so, in all the cases reported, introductions had first been made by a senior member of the university administration.

Table 13.1 summarises the sample population of university firms and Table 13.2 gives information on the founders of these companies.

Table 13.1 Sample of 31 university-linked firms

Linked to University	Number	%
QUB (Belfast)	10	32.2
TCD (Dublin)	4	12.9
UCL (London)	4	12.9
Manchester	2	6.5
UMIST	5	16.1
Leeds	4	12.9
Linköping (Sweden)	2	6.5
Total	31	100.0

Table 13.2 Founders of university-linked firms

University staff	20	65%
Postgraduate student(s)	4	13%
Existing company	4	13%
Private entrepreneur(s)	3	10%
University holding company	17	55%

In all cases but one, the founder(s) were still active participants in the companies, in executive roles and/or members of the companies' Boards of Directors.

Methodology

An approach was made through the relevant industrial liaison professional at each university with a request for introductions to the chief executive or leading personality in a sample of their university companies. Face-to-face interviews were then held during which a questionnaire was completed. The main sections of the questionnaire sought information about:

- Company formation, age and present status;
- Products/Services and competitive edge;
- Inputs & Sources of Supply;
- Markets and Marketing;
- Staffing and Organisation;
- Company Origins and Links with the university;
- Company Performance.

Finally, to compare the views of the company chief executives about the most important factors affecting their companies with those of the university administrators, they were invited to score the same set of factors as had been included during the earlier interviews with university staff, (see Chapter 8, Figures 8.1 and 8.2 and Tables 8.1 and 8.2). The interviews lasted from 35 minutes to 90 minutes and averaged just less than one hour. All the interviews were held between December 1992 and July 1993.

Products, Services and Industrial Sectors

The products and services offered by the sample companies included:

> Printing & text applications of information technology; Software; Marine environmental consultancy; Medical diagnostics; Thin film surface coatings, opto-electronics; Special purpose chemicals fluorescence; Electronic components; CAD system software distribution, IT; Educational publishing; Gas turbine development and manufacture; Food/nutrition consultancy; Instrumentation of plastics process technology; Scientific instruments/light-based technology; Audio components/digital electronics; Business software products and consultancy; Scientific instruments; Biomedical instrumentation; Vaccines; Mapping; Medical screening; Environmental monitoring; Materials technology; Biotechnology.

Twenty of these companies were predominantly product-based, ten provided mainly knowledge-based services or contract research and consultancy and about half offered a mixture of products and services.

Age of Companies

Table 13.3 shows the age profile of the sample. The longest established company was formed in 1983 and the newest in 1993.

Table 13.3 Age distribution of university-linked firms

Age at date of interview	Firms	%
<1 year	3	9.7
1-3 years	9	29.0
>3 -5 years	9	29.0
>5-10 years	10	32.3

Almost 40 per cent of the companies had been in business for three years or less at the time of interview and slightly less than one-third had been in existence for more than five years. Several of the enterprises had been operating for a number of years before company formation as university-based industrial units and centres and did not begin from a standing-start when they became companies. The three most recently formed companies were only just getting started and had not made significant sales at the time of interview.

Competition

To assess the source of their competition the companies were asked to identify their main competitors by location. All identified main competitors from outside their own region. Table 13.4 shows for example that 86 per cent of the 28 firms which responded had main competitors in Great Britain (but which were generally outside of their local region) and over 70 per cent also faced competition from overseas. These firms viewed themselves as having to compete with companies worldwide. As will be seen, this is in contrast to the results from a group of low-technology firms interviewed later, see Chapter 14.

Table 13.4 Location of competitors of university-linked firms

Country	Firms	%
Great Britain	24	86%
Northern Ireland	2	7%
ROI	2	7%
Other EC	18	64%
USA	19	71%
Other	10	43%

Markets and Marketing

Twenty-nine firms responded to this question which asked for a breakdown of the destination of last year's sales. The two companies which declined felt they had not been trading long enough to estimate their position. Of the others, 20 firms, (69 per cent) had some local turnover - defined as being within a 20 mile radius - and averaging 27 per cent of turnover.

Overall these 29 firms 'exported' 82 per cent of their products from their local region. Twenty-five (86 per cent) did more than half of their business beyond their region and 21 (72 per cent) exported more than four-fifths of their turnover. The ultimate destinations of the products of some companies were much farther afield than the results at first seemed to indicate because several were selling to associated companies in GB for onward distribution overseas. The overall picture which emerged was of a group of companies serving world markets and acutely aware of international competition.

Table 13.5 shows the marketing methods used by these companies. Nineteen companies used more than one method of marketing, two used four methods and one used five different methods. The question simply asked which methods were used and not about their relative frequency, however some companies relied almost exclusively on their own marketing. Many university companies readily admit to having limited expertise in marketing and yet 93 per cent of the firms in this sample engaged in some direct marketing of their products. Given the almost universally high importance given to marketing by those interviewed, see Tables 8.1 and 8.2 and Figures 8.1 and 8.2, the wisdom of this approach should at least be questioned. This aspect of the companies' activity would benefit from further research.

Table 13.5 Marketing methods used by 31 university-linked firms

Marketing methods	Firms	%
Direct by the company	27	93%
Via parent or associated company	6	21%
Via agents	14	48%
Mail order	4	14%
Other	5	17%

New Products and Innovation

A series of questions was asked to explore each company's attitude to innovation, new product development and the stage each of its products had reached in its life-cycle. Companies were asked specifically whether any new products had been introduced during the last three years. They were asked to look back three years and then ahead three years to assess the changes in their product portfolio. Twenty-one companies, 68 per cent of the sample, had introduced at least one new product in the previous year and 27 companies (87 per cent) had done so during the last three years. It is to be expected that recently formed companies would have introduced new products but even amongst the 19 longest established firms, i.e. those which had been in business for more than three years, 12 (63 per cent) had introduced new products during the previous year and 15 (79 per cent) had done so during the last three years.

Product Life Cycle

As a further indication of companies' rates of innovation Table 13.6 sets out the number and percentage of firms in the sample having products at various stages of growth and maturity. (The total percentage exceeds 100 because, typically, companies have products at different degrees of maturity.)

Table 13.6   Number of firms with products at various stages of maturity

Stage of maturity	Firms	%
Embryonic	17	55%
Demand growing rapidly	18	58%
Demand growing more slowly	14	45%
Demand on a plateau	8	26%
Demand declining	5	16%

Again, a high proportion of recently formed companies are likely to have products at early stages of their life cycles. Table 13.7, however, shows that a considerable proportion of the longer established firms in the sample also had new and embryonic products as well as their share of more mature lines.

Table 13.7 Older firms having products at various stages of maturity

Stage of maturity	Firms	%
Embryonic	8	42%
Demand growing rapidly	13	68%
Demand growing more slowly	8	42%
Demand on a plateau	6	32%
Demand declining	4	21%

Tables 13.8 and 13.9 display the total number and percentage of products at various stages of growth and maturity, firstly for all firms in the sample and secondly for the 19 longest established firms.

Table 13.8 Number and percentage of products in all firms sampled

Stage of maturity	Products	%
Embryonic	32	30%
Demand growing rapidly	30	29%
Demand growing more slowly	22	21%
Demand on a plateau	12	11%
Demand declining	9	9%
Total products in 31 companies	105	100%

Table 13.9 Number and percentage of products in oldest firms

Stage of maturity	Products	%
Embryonic	10	19%
Demand growing rapidly	19	36%
Demand growing more slowly	12	23%
Demand on a plateau	7	13%
Demand declining	5	9%
Total products in 19 companies	53	100%

There is excellent correlation between the percentages of all firms and products i.e. Tables 13.6 and 13.8 (a correlation coefficient of 0.98) and a less good but still reasonable correlation coefficient of 0.78 when the oldest firms only are included i.e. Tables 13.7 and 13.9.

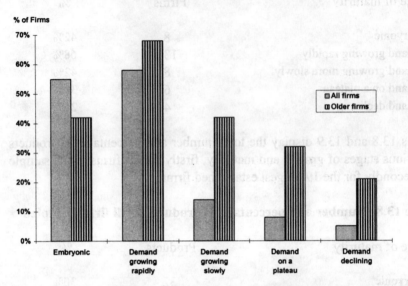

Figure 13.1 Percentage of firms with products at various stages of maturity

Figure 13.1 displays the profile of firms having products at various stages of maturity and Figure 13.2 shows the percentage of products for all firms at various stages. In each case the charts show the picture when all companies in the sample are included and that when only those which have been in business for more than three years are included.

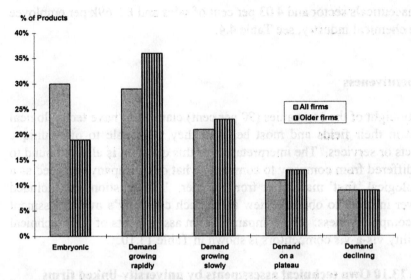

Figure 13.2 Percentage of products at various stages of maturity

As would be expected, in both Figures 13.1 and 13.2 the proportion of embryonic products decreases when the newer firms are removed from the sample, however the overall picture is of a group of companies whose product profile is skewed towards the embryonic and young end of the spectrum, suggesting that they have a commitment to innovation and new product development.

R&D Expenditure

Closely associated with technological innovation is annual expenditure on R&D. This ranged from 'very little' to £1.5m with a mean of £181k. Expressed as a percentage of total revenue this translates to a maximum R&D expenditure equal to 95 per cent of sales and an average of 24 per cent of sales. R&D Expenditure per employee ranged from £1k to £60k and averaged £10.9k. By almost any criterion therefore these companies would be categorised as being high-technology-based. It is notable that according to Company Reporting (1993), the R&D expenditure among the top 200 companies of the UK's two highest spending industrial sectors were less than 7.68 per cent of sales and £5.89k per employee by the pharmaceuticals sector and 4.03 per cent of sales and £ 3.69k per employee by the chemical industry, see Table 4.4.

Competitiveness

Twenty-eight of the companies (90 per cent) claimed to have technological 'firsts' in their fields and most believed they were able to offer unique products or services. The interpretation of this question is almost bound to have differed from company to company, what one company regarded as a technological 'first' may differ from another. The question was included however in order to obtain a view about each company's own assessment of its competitiveness. The companies' own assessments of their technical capability vis-à-vis competitors is shown in Table 13.10.

Table 13.10 Own technical assessments by university-linked firms

The leader	12	39%
In top ten per cent	15	48%
In top quartile	3	10%
Median	1	3%
Total	31	100%

These answers indicate a high degree of confidence on the part of those interviewed in their products and their competitive edge. A more objective indication about technical leadership was obtained from the answers to another question which asked whether the company had access to unique patents or other intellectual property. Twenty-six companies (84 per cent) claimed to have access to one or more patents, unique know-how or other intellectual property. Of these, 19 obtained this from their founding university, five owned the IPR themselves and two licensed-in the technology from an overseas company.

University Links

The high value placed by companies on maintaining links with their 'parent' university is unsurprising amongst those which still rely on the university as a source of technical assistance and access to resources and equipment. Even those firms which had matured to the point of being relatively independent still valued their university connections, not least as a source of appropriately skilled personnel, Table 13.11.

Table 13.11 Value of links with 'parent' university

University Links	Firms	%
Close	27	87%
Remote	3	10%
Vital	8	26%
Very useful	15	48%
Useful	4	20%
Fairly useful	7	23%
Neutral	1	3%

Staffing and Numbers Employed

Seventy-two per cent of those employed were graduates or postgraduates. This is an extremely high percentage even for high-technology-based firms and appears to be one of the distinguishing features of university spin-outs, at least of those at this stage in their development. It is about three times as high as the percentages of graduates in the non-university high-technology companies interviewed later, Chapter 14.

Table 13.12 Staff qualifications numbers, percentages and average per firm

	Number employed	%	Average per firm
Graduates	216	45	7.0
Postgraduates (higher degrees)	127	27	4.1
HND/HNC etc	30	6	1.0
Administrative/clerical	44	9	1.4
Others	61	13	2.0
Total current year (year of interview)	478	100	15.4

Staff Shareholdings

In 22 companies one or more members of staff had shareholdings, ranging from 0.5 per cent to 65 per cent of the equity and in most cases they were also on the Boards of Directors. In 14 instances the shareholdings were acquired by personal investment, four were gifts and 13 were in exchange for expertise. These stakeholdings result in a very high degree of commitment to the university companies and motivation to make them successful although not, as has been seen earlier solely or even mainly for financial gain.

Turnover

Twenty-seven companies supplied historical data on turnover, Table 13.13.

Table 13.13 Sales revenue (turnover) of 27 university-linked firms

(Current year is date of interview)	Last year	Current year	Next year (est.)
Total Turnover (Revenue)	£17,933,000	£20,930,000	£25,550,000
Average Turnover	£664,000	£775,000	£911,000
Factor (current year =100)	86	100	117.5

Most of these firms were anticipating increased growth next year, average 17.5 per cent and the four more recently formed companies, whose data was not included, were expecting considerably more. Given that the weighted average age of these firms was five years this is quite rapid but not spectacular growth. It must not be overlooked that this is anticipated as distinct from actual growth and the estimates were made at a time when the UK economy was in recession. Employment growth was not recorded but in the QUBIS Limited case-study (Chapter 12) growth of employment was shown to correlate well with growth of turnover. In the QUBIS Limited companies, ten of which are included in the sample of 27 which provided sales data, the growth of turnover between years four and five was 30 per cent and employment growth was 32 per cent. The IRDAC study of 12 firms quoted a growth of 30.8 per cent between 1983 and 1985 (Rothwell and Dodgson, 1987).

Materials Cost and Added Value

Twenty-nine companies estimated their material costs, ranging from less than 5 per cent to 58 per cent of sales value. The mean value of 18 per cent of sales value, gives a crude estimate for added value of 82 per cent of sales. The added value per employee averaged £41k (sample of 24 firms). For comparison the mean, median and upper quartile values for all the companies included in the Performance Benchmarks for Developing Firms[2], were £28.9k, £24.1k and £35.4k (NIERC, 1996). The added value per employee of 12 (i.e. 48 per cent) of the 25 university companies which provided data exceeded the upper quartile in the Benchmark firms and 16 firms (64 per cent) exceeded the median value, a further two firms were within 2 per cent of the median Benchmark.

Profit Before Tax and Equity Subscribed

Those companies which provided information on profit before tax as a percentage of sales reported average figures as in Table 13.14.

Table 13.14 Profitability of university-linked firms

(Current year is date of response)	Last year 19 firms	Current year 23 firms	Next year (est.) 24 firms
Profit before tax (PBT)	9%	12.3%	14%
Factor (current year =100)	73	100	114

For comparison the median and upper profitability figures in the NIERC Benchmarks were 5.4 per cent and ten per cent respectively. Of the 23 companies which provided profitability figures for the current year 13 (56 per cent) exceeded the Benchmark median and 12 (52 per cent) had profits above the Benchmark upper quartile (NIERC, 1996). One company reported a loss and two others reported a break-even position for the current year. Of those which did not respond to this question, two firms were embryonic and had less than a full year's sales at the time of interview and another which was preparing for a stockmarket launch did not disclose its profits.

Twenty-six companies estimated the equity subscribed, which varied from less than £1k to £7.5m. The mean value was £410k but after excluding one company which had just been floated on the stock exchange for £7.5m, the average equity subscribed was £126k.

Because many of these companies were started with very small sums invested - one firm claims to have had only £1,000 initial investment - return on investment calculations result in unrealistically high figures. On the other hand there have been fairly high uncosted investments of university staff time and also in some cases uncosted access to university facilities. Without more detailed information than was available calculations of return on investment are fairly meaningless.

Contribution to Local Economy

To assess the impact which the companies had on their local economies through purchases of equipment and materials they were asked to estimate their sources of supply for capital items and for consumables i.e. non-capital items (excluding rent, rates, utility services etc) The results are presented in Tables 13.15 and 13.16.

Table 13.15 Sourcing of capital items by university-linked firms

Source for capital items by university-linked firms	Average % of expenditure
Local, 20 mile radius	27
Within 100 mile radius	12
Other UK	48
Republic of Ireland (ROI)	4
Elsewhere	9

Most capital items were purchased outside the local region and the majority of routine purchases were made locally. Salaries were almost universally also paid locally.

Table 13.16 Sourcing non-capital items by university-linked firms

Source for non-capital items by university-linked firms	Average % of expenditure
Local, 20 mile radius	71
Within 100 mile radius	8
Other GB mainland	17
Republic of Ireland (ROI)	4
Elsewhere	1

Taken together with the fact that most of the sales turnover comes from outside the local area it is clear that the economic impact of these companies on their local economies will be maximised. Companies were asked to assess the means by which they contributed to their local economy. Table 13.17 shows the number of companies which contributed under each heading.

Table 13.17 Perceived contribution to local economy by university-linked firms

Impact	Firms	%
Introducing a new industry	3	10
Introducing new technology to existing industry	22	71
Creating new jobs	29	93
Employing graduates	25	81
Contributing to exports from the region	15	48
Supporting local suppliers and other ways	12	39

Summary

The results of this sample of 31 university-linked firms add credence to many of the hypotheses made early in this Chapter. Specifically they show:

- The companies are predominately export orientated and view their competitors as being international rather than local;
- Despite an expressed concern about marketing and an admission that universities are not strong in this area, 93 per cent of the companies engaged in direct marketing of their products;
- The companies showed evidence of innovation in that two-thirds of even the longer established firms had introduced new products in the previous year. Furthermore their annual R&D expenditure was also higher than most UK companies at an average of 24 per cent of sales value. This was backed up by statistics which showed a good spread of products at various stages of their life-cycles and high proportions of embryonic and relatively immature products;
- Most companies believed they were competitive and 87 per cent believed that their technical capability was in the top ten per cent or better in their field;

- Links with their parent university were strong and valued;
- Seventy-two per cent of those employed in the companies were either graduates or postgraduates and in 22 out of the 31 companies one or more staff members had shareholdings;
- Average turnover was £775k and expected growth of 27 companies who provided sales data was expected to average 17.5 per cent in the following year;
- Added value averaged 82 per cent of sales and added value per employee was £41k;
- Average profit before tax, based on a small sample, averaged 12.3 per cent of sales and was forecast to rise to 14 per cent in the following year;
- The companies bought most capital items outside their local region but non-capital items were almost always sourced locally. Salaries were almost entirely paid locally so the impact of these companies on their local region is maximised.

Notes

1. Although not included in the sample of companies in the survey, discussions about the characteristics of spin-out firms in Europe were held during visits with Rikard Stankiewicz at University of Lund and with Dick van Barneveld and Jan Kobus about the TOPS and TOS programmes for new company creation at the University of Twente in The Netherlands.
2. The Performance Benchmarks for Developing Firms are based on a sample of 700 developing manufacturing firms in Ireland. A developing firms has between 10 and 100 employees and a desire to achieve rapid growth in profitability and/or sales. The survey is part of the Competitive Analysis Model (CAM) project, a joint initiative between LEDU and IDB in Northern Ireland and Forbairt in the Republic of Ireland. The project is conducted by staff at the Northern Ireland Economic Research Centre (NIERC).

14 Questionnaires to Small Non-University Firms

Introduction

This Chapter draws on some of the results of a three year university Outreach Programme to SMEs in Northern Ireland carried out under the authors' supervision and completed in June 1995. Amongst the objectives of this programme which were achieved through an extensive programme of carefully pre-planned company visits to a sample of SMEs across a range of industrial sectors were:

- To create and maintain a greater 'presence' for Queen's University in the local, small firm sector;
- To listen to the problems being faced by smaller firms;
- To obtain a better understanding of the smaller firm sector;
- To learn of their views and perceptions of the University;
- To make known to them the services which the University can offer.

In the present context the objectives of using data from the Outreach Programme were firstly, to enable differences between high-technology (HT) and low-technology (LT) based firms to be assessed and secondly, to provide a group of non-university based small firms for comparison with the university firms described in the previous Chapter. In anticipation of this comparison, the questionnaire used for the outreach programme was designed to include a series of questions identical to those asked of university firms.

This Chapter, like the previous one, is mainly confined to reporting the results of the questionnaires and interviews. Comparisons and contrasts between the LT and HT firms of the non-university firms are discussed but comparisons with the university sample and interpretation of the data will be made in Chapter 15.

The Sample

The small firm agency, Local Enterprise Development Unit (LEDU), which was approached for assistance in the selection of firms to include in the sample, proposed 156 companies which were judged to be 'preparing for growth'. Although these companies fell fairly naturally into 11 industrial sectors, they are not necessarily representative of the small firms sector in Northern Ireland. For example one sector categorised by LEDU as miscellaneous manufacturing was in fact a 'catch-all' for a number of small firms including some in chemicals, paper, plastics, printing, packaging, clothing and textiles. The LEDU nominated firms were augmented by others chosen at random from the Industrial Development Board for Northern Ireland (IDB) trade directory. The criteria for selection were, size - generally less than 50 employees (the average number employed was 32); local - Northern Ireland based and owned; and independent - not part of a larger group.

Methodology

For each company, the most senior person, usually the owner/manager, was identified and contacted either by telephone or by letter to explain the purpose of the programme and to request a meeting. An undertaking to respect confidentiality was given and this contributed to the willingness with which the questions were subsequently answered. The initial approach to each company, which was refined as the programme progressed, was an important contributory factor to the success of the subsequent meetings. The response was generally excellent, once initial suspicions had been allayed, and about 95 per cent of those approached agreed to a meeting.

Meetings ranged from 45 minutes to three hours and averaged about 75 minutes. Efforts were made to keep the interviews as informal as possible and the interviewees were encouraged to talk about their businesses in general. The meetings were structured to ensure that, following a free-ranging discussion, a series of specific questions was asked, many of which were identical to those used for the university-based companies. The questions were adapted slightly to suit the circumstances of each firm and not all questions were entirely appropriate in all cases.

Since not all questions were asked, and in some cases those which were asked were not fully answered, the numbers responding varied from question to question. This is made clear in the tables of results below.

Answers were recorded either during the interviews or very shortly afterwards. Efforts were made to capture the 'tone' of the replies and the ambience of the meetings as well as the factual responses. Most of the interviews were held between first quarter 1992 and the end of 1994, a period during which the UK and the local Northern Ireland economy were emerging from recession.

More than 400 firms were visited. Some were excluded from the analysis either because they provided inadequate data or because when visited they failed to meet one or more of the criteria determined for inclusion in the outreach programme - frequently because they were too large. Results from 376 SMEs are included in the analysis which follows. For the purpose of the comparison with university companies the firms were subsequently categorised as being either High-Technology based (HT) or Low-Technology based (LT). Initially this was a subjective assessment made immediately following each visit and before the results were analysed. In fact it proved to be quite easy to categorise most of the companies in this way. The initial selections were then reassessed after the data for all the companies had been compiled. Those categorised as being in the high-technology sector were then examined for the proportions of science, engineering or technology graduates employed. As a first pass, to qualify for inclusion in the HT group, firms had to have at least 20 per cent graduates. Twenty-eight companies met this criterion and a further 13 were admitted as high-technology-based firms on other criteria related to the nature of the business and its dependence on technology. Overall 25 per cent of employees in the final HT group were graduates compared with three per cent in the LT companies. The final categorisation was 336 low-technology and 40 high-technology firms.

Industrial Sectors and Technologies Covered

Table 14.1 lists the 11 sectors into which the sample companies were categorised.

Table 14.1 Number of companies by industrial sector/products

Sector/products	HT firms	LT firms	Total
Construction engineering	0	69	69
Software	19	2	21
Vehicles & accessories	0	14	14
Miscellaneous manufacturing[a]	7	72	79
Precision engineering	0	27	27
Materials handling & hydraulics	0	35	35
Metal fabrication	0	17	17
Electrics and electronics	8	21	29
Food processing	4	49	53
Miscellaneous engineering[b]	2	13	15
Agricultural engineering	0	17	17
Total	40	336	376

Notes [a] The products of the HT firms in the Miscellaneous manufacturing category were - Medical products; Powder coatings and paints; Textile fabrics; Textile finishers (three firms); Chemical flocculants.

[b] The HT firms in the Miscellaneous engineering category produced - Process plant design for petro-chemicals, food and chemical plants; Remote instrumentation and process controllers.

Location

Table 14.2 shows the geographical location of the companies within Northern Ireland. Note the clustering of technology-based (HT) firms in the Greater Belfast area, mainly as a result of the preponderance of software firms in the HT category, 74 per cent of which were located in or around Belfast.

Table 14.2 Geographical distribution in Northern Ireland

Location in NI	H T	LT	All
Greater Belfast	59%	30%	34%
North East	24%	27%	27%
South West	2%	14%	13%
South East	0%	18%	16%
North West	15%	11%	11%

Age of Companies

Table 14.3 shows the age distribution of firms, defined as the number of years which the firm had been in business at the date of interview. Almost three-quarters were over ten years old, although there were quite large sectoral differences.

Table 14.3 Age of firms by sector, all firms

Sector / products	< 1 year	1-5 years	5-10 years	>10 years	Firms
Agricultural engineering	0%	0%	0%	100%	11
Construction engineering	0%	10%	30%	59%	69
Electrical/electronic engineering	0%	7%	14%	79%	14
Food processing	0%	6%	8%	86%	49
Materials handling & hydraulics	0%	11%	11%	78%	18
Metal fabrication	n.a.	n.a.	n.a.	n.a.	2
Miscellaneous. engineering	9%	9%	18%	64%	11
Miscellaneous manufacturing	1%	6%	14%	78%	77
Precision engineering	0%	18%	18%	64%	11
Software	10%	29%	33%	29%	21
Vehicles & accessories	0%	0%	7%	93%	14
Overall	1%	9%	18%	72%	297

The longest established sectors were agricultural engineering, vehicles and accessories and food. By contrast, 29 per cent of software companies were less than ten years old and as many as 39 per cent of these had been in existence for only five years. There were, proportionately, almost three times as many firms under five years old in the HT sector as the LT sector. Half of those in the miscellaneous engineering, 31 per cent in software and 30 per cent in food processing being less than five years old, Table 14.4.

Table 14.4 Age distribution of non university-linked firms

Age at date of interview	HT n = 36	LT n = 260	All n = 296
< 1 year	6%	1%	1%
>1- 5 years	19%	8%	9%
>5-10 years	22%	17%	17%
>10 years	53%	75%	72%

Products and Services

Table 14.5 shows the products and services offered. Although the firms selected were primarily in manufacturing, it transpired that many also provided services of various kinds and some were agents or factors for products made by others. As a result of this diversity the percentages in Table 14.5 exceed 100 per cent.

Table 14.5 Activities of companies in sample

	HT firms		LT firms		All firms	
Manufactures own product	37	90%	259	77%	296	79%
Manufactures to sub-contract	9	22%	115	34%	124	33%
Service provider	19	46%	81	24%	100	27%
Sales or distribution	8	20%	45	13%	53	14%

Competition

Companies were asked to identify their main competitors and their location (Table 14.6). Almost the same percentage of firms had competitors outside Northern Ireland (64 per cent) as inside (62 per cent). However there were large sectoral differences, only the agricultural engineering, food and software sectors saw themselves as competing internationally to any great extent. Many of the businesses appeared deliberately to position themselves so as not to compete with national or international businesses or with GB competitors. Hence the importance of sub-contract work or business with a large service element, which is 'import proof' but also difficult to export.

Table 14.6 Percentage of firms with competitors in regions indicated

Sectors	NI	ROI	GB	EU	Overseas
Agricultural engineering	65%	41%	53%	0%	6%
Construction engineering	87%	9%	25%	3%	0%
Electrical/electronic engineering	75%	5%	25%	5%	5%
Food processing	73%	38%	49%	19%	14%
Materials handling & hydraulics	64%	27%	42%	15%	6%
Metal fabrication	76%	18%	18%	12%	0%
Miscellaneous. engineering	71%	14%	29%	7%	14%
Miscellaneous manufacturing	62%	18%	45%	9%	15%
Precision engineering	86%	5%	9%	5%	5%
Software	68%	21%	53%	5%	21%
Vehicles & accessories	75%	8%	50%	8%	17%
Overall	62%	16%	31%	7%	7%

There were marked differences between the HT and LT firms. Half of the high-technology firms had major competitors in Northern Ireland but even more also saw themselves as competing with companies in GB and over 20 per cent had major competitors overseas.

Table 14.7 Location of main competitors

Country	HT	LT	All
Great Britain	51%	29%	31%
Northern Ireland	54%	63%	62%
ROI	17%	15%	16%
Other EC	5%	7%	7%
Overseas	20%	6%	7%

This is particularly marked amongst software firms and in that part of the electronic/electrical sector which is not related to construction and house wiring. By contrast the LT group was more preoccupied with local competition, 29 per cent had serious competitors from GB and only 6 per cent from firms overseas, Table 14.7.

Markets

Although in both groups some business was done locally, a much greater percentage of business was done outside Northern Ireland by the technology-based firms (Table 14.8).

Table 14.8 Percentages of sales outside Northern Ireland

	% of firms having some sales outside NI	% of firms with 50% or more sales outside NI	% of firms with 80% or more sales outside NI
HT firms	85%	54%	38%
LT firms	74%	25%	12%

New Products and Innovation

One indicator of the commitment to innovation is the extent to which new products are introduced. Overall, less than 50 per cent of the companies in the sample had introduced a new product in the previous two years. This is disturbing as it seems to indicate a reluctance on the part of most SMEs not only to engage in Research and Development in its broadest sense but to keep abreast of new developments and products introduced by competitors. Almost certainly this is because such a high proportion of, mainly LT, firms faced only local competition. The contrast between the HT and LT groups was stark. Twenty-five, 61 per cent, of the high-technology-based firms, had introduced at least one new product in the previous year and 95 per cent had done so during the last three years. The corresponding figures for the low-technology group were much lower - only 31 per cent of which had introduced a new product during the past year and 52 per cent had done so in the last three years, see Table 14.9.

Table 14.9 Companies which had introduced new products

New product introduced	HT firms		LT firms		All firms	
Last year	25	61%	105	31%	130	34%
or 2 years ago	10	24%	48	14%	58	15%
or 3 years ago	4	10%	25	7%	29	8%

Almost half of the owner/managers interviewed were unable to estimate their company's expenditure on R&D. In some cases this was because such expenditure, especially that on product development, was not specifically separated from general operating costs but also, one suspects, because in many cases no R&D was taking place. One hundred and thirty-five companies in the LT group estimated their expenditure on R&D as a percentage of sales, which ranged by sector from zero to 17 per cent but averaged only 1.6 per cent. Thirty-one of the 40 HT firms were able to make an assessment of their R&D expenditure. This ranged from zero to 40 per cent and averaged 4 per cent of sales. The average per sector varied from a high of 4.6 per cent of sales for electrical/electronics down to zero reported by two companies in miscellaneous engineering.

It is almost certain that R&D expenditure is not being accurately identified as such in some of these responses. Were it not for other indicators - such as the nature of the companies' products and the proportions at early stages in their life cycles, the markets into which these firms are selling and proportion of graduates and those with higher degrees employed - one would have to question whether these firms were properly classified as being high-technology based. The tendency for smaller firms to understate their R&D was noted by Kleinknecht (1987), who concluded that the underestimation was because of the informality of the R&D process, much of which in small firms was to do with product development which does not fulfil the definitions of research and development used by OECD and other bodies engaged in international surveys.

Product Life Cycle

Interviewees were asked to indicate the stages which their products had reached in their life cycles[1]. Table 14.10 shows the number and percentage of firms in the sample having products at various stages of growth and maturity. The total percentage exceeds 100 per cent because companies typically have products at different stages of maturity. Although the HT firms had, proportionately, more than twice as many embryonic and rapidly growing products, both groups had a considerable number of mature and declining products.

Table 14.10 Firms with one or more products at various stages of maturity

Maturity of products	HT firms 41 firms		LT firms 282 firms		All firms 323 firms	
Embryonic	16	39%	51	15%	67	18%
Demand growing rapidly	23	56%	79	24%	102	27%
Demand growing more slowly	28	68%	158	47%	186	49%
Demand on a plateau	16	39%	150	45%	166	44%
Demand declining	7	17%	30	9%	37	10%

Table 14.11 sets out the percentage of products at each stage of the life cycle given for each group. Forty-two per cent of products in the HT sector were either embryonic or at a rapidly growing stage of their life cycle compared with only 25 per cent for the LT group.

Table 14.11 Number of products at various stages of their life cycles

Maturity of products	Products of HT firms		Products of LT firms		Products of All firms	
Embryonic	18	15%	65	9%	83	10%
Demand growing rapidly	33	27%	112	16%	145	18%
Demand growing more slowly	41	34%	252	37%	293	36%
Demand declining	7	6%	30	4%	37	5%
Demand on a plateau	23	19%	230	33%	253	31%
Total	122	100%	689	100%	811	100%

Another source of comparison is the Benchmarks for 'Contribution to Sales of New, Improved and Unchanged Products' (NIERC, 1996). This showed contributions of 13.6 per cent from new products, 16.3 per cent from improved products and 70.2 per cent from unchanged products. If the embryonic products and those for which demand is growing rapidly, in Table 14.11, were regarded as being roughly equivalent to new and improved products they would account for 28 per cent of all 811 products in the sample. Since nothing is known about the relative contribution to income from these products, the figures are not strictly comparable with the Benchmark data but the figure of 28 per cent is close to the 30 per cent for new and improved products in the Benchmarks.

Although the number of HT firms is too small to make an accurate analysis of the sectoral differences, the overall picture displays some interesting variations between the sectors. Tables 14.12 and 14.13 correspond to Tables 14.10 and 14.11 for the sample companies as a whole, by sector. Table 14.12 shows the sectoral analysis of product life cycles by percentage of companies. This table is ranked by the percentage of firms having one or more embryonic products and once again the same sectors most exposed to international competition i.e. software, agricultural engineering, electronic engineering and food reported the greatest proportion of embryonic products.

Table 14.12 Firms with products at various stages of maturity

Sector (number of firms in brackets)	Embry-onic %	Demand growing rapidly %	Growing more slowly %	Mature on a plateau %	Demand declining %
Software (21)	48	76	48	48	24
Agricultural engineering (15)	27	53	67	53	13
Electronic engineering (23)	26	22	74	39	17
Food (34)	24	62	62	59	15
Miscellaneous engineering (14)	21	29	64	21	21
Miscellaneous manufacturing (71)	21	31	68	48	7
Metal fabrication (17)	18	18	35	71	6
Material handling / hydraulics(28)	18	21	54	54	4
Construction engineering (68)	15	22	46	59	7
Vehicles / accessories (13)	14	nil	62	38	31
Precision engineering (18)	6	11	56	56	11
Total % (322 firms)	21	32	58	52	11

For comparison with Table 14.12 which shows the number and percentage of companies with products at various stages, Table 14.13 shows the percentage of products at each stage of the life cycle given for each sector, e.g. 16 per cent of the products in the electrical & electronic engineering sector were at the embryonic stage compared to three per cent of the products in the precision engineering sector. There is strong correlation between the data in Tables 14.12 and 14.13, with the exception of the vehicles sector where the correlation coefficient is 0.73. In all other sectors the correlation coefficient is greater than 0.95.

Table 14.13 Products at various stages of maturity, as percentage of all products

Sector	Embry -onic %	Demand growing rapidly %	Growing more slowly %	Mature on a plateau %	Demand declining %
Electrical/electronic engineering	16	11	44	20	9
Software	15	35	20	24	7
Agricultural engineering	7	26	40	23	4
Food	11	27	33	25	4
Miscellaneous engineering	12	16	48	12	12
Miscellaneous manufacturing	12	16	42	27	3
Metal fabrication	10	10	28	48	3
Material handling / hydraulics	10	12	42	35	2
Construction engineering	6	14	30	47	3
Vehicles / accessories	8	nil	52	24	16
Precision engineering	3	6	41	44	6
Total % (811 products)	10	18	36	31	5

Competitive Edge

Twenty-one HT companies (48 per cent) claimed to have technological 'firsts' in their fields and a further 36 per cent, believed they had other significant technical advantages over their competitors. The corresponding figures for the LT firms were 90 companies (30 per cent) with technological firsts and an additional 22 per cent claiming to have other advantages. The companies' own assessments of their technical capability vis-à-vis competitors is shown in Table 14.14

Table 14.14 Firm's own technical assessments

Firm's own technical assessment	HT 33 firms	LT 264 firms	All 297 firms
The leader	27%	31%	31%
In top 10 per cent	48%	44%	45%
In top quartile	9%	5%	6%
Median	6%	17%	15%
< Median	9%	3%	3%

There are some reservations about these answers because although the question asked for a technical assessment, the replies were frequently qualified and a number of companies claimed to be leaders 'in what we do'. It was apparent that many respondents had interpreted the question as going beyond purely technical leadership. The results therefore are more indicative of the perceptions of the respondents than an objective assessment of technical advantage.

Staffing and Numbers Employed

The numbers and percentages of graduates varied by sector. From Table 14.15 it is clear that the software sector was by far the largest graduate employer in this group.

Table 14.15 Graduates and those with higher degrees by sector

Sector	Firms	Total Full-time staff	Graduates & higher	Percentage graduates
Agricultural engineering	17	583	9	1.54%
Construction engineering	69	2,114	93	4.40%
Electrical/ electronic engineering	28	1,066	74	6.94%
Food	31	1,325	44	3.32%
Materials handling/hydraulics	34	803	29	3.61%
Metal fabrication	16	431	4	0.93%
Miscellaneous engineering	15	340	31	9.12%
Miscellaneous manufacturing	79	3,461	114	3.29%
Precision engineering	27	827	18	2.18%
Software	21	511	199	38.84%
Vehicles/ accessories	13	395	2	0.51%
Total	350	11,856	617	5.20%

The overall proportion of graduates and those with higher qualifications employed, 5.2 per cent, is rather lower than would have been expected when compared with data from the NIERC Product Development Survey database for manufacturing firms employing up to 100 which shows a mean for Northern Ireland of 6.6 per cent.

Table 14.16 Staff qualifications numbers and percentage of total employed

Staff categories	HT firms		LT firms		All firms	
Graduates and postgraduates	291	25%	326	3%	617	5%
Other technical qualifications	62	5%	328	3%	390	3%
Other staff	806	70%	10,180	94%	10,986	92%
Total employed	1159	100%	10,834	100%	11,993	100%

It comes as no surprise that the greatest contrast between the HT and LT groups is in the number of graduates employed, see Table 14.16, as that was the main criterion used to separate and categorise the two groups. In 26 per cent of the HT firms more than 40 per cent of the staff had graduate or higher qualifications.

Also significant were the underlying attitudes to graduate recruitment and the preconceptions which many managers of the firms in the sample held. There was a widespread perception that graduates lack practical skills. This, and the high cost of training deterred many small firms from recruiting new graduates. This point was most frequently expressed as a complaint that it took longer to train graduates than those who had been brought into the firm directly from school. The expectation amongst many owners and managers of the SMEs interviewed - particularly those in the more traditional industries, most of whom had not themselves had the benefit of university education, - is that graduates should be able to make a more immediate contribution to the business. The fact that, in the longer run, graduates are eventually able to contribute more by virtue of their greater knowledge and ability often is not appreciated.

One effect of this is that a number of these smaller company employers were reluctant to recruit recent graduates and yet many felt that they could not afford to pay the higher wages associated with experienced graduates. Graduate employment therefore was seen as an unrealistic option by many of the businesses because of perceived high wage costs, high mobility, high training costs and higher expectations on the part of graduate employees. This is unfortunate given the correlation between a company's innovative performance and the employment of graduates, e.g. compare the percentages of graduates employed and the percentage of firms with embryonic products in each sector (Tables 14.15 and 14.12), the correlation coefficient is 0.85.

Sales and Added Value

Many of the companies were reluctant to discuss sales and even fewer were prepared to reveal their profitability. However, enough were willing to estimate the material content as a percentage of sales to enable estimates of added value and relative productivity of labour i.e. added value per employee, to be made.

Table 14.17 Sales and added value (average figure) by sector

Sector	Sales £ million	Added value as percentage sales	Added value per Employee £ thousand
Agricultural engineering	2.0	57%	33
Construction engineering	2.4	68%	50
Electrical/electronic engineering	1.3	64%	23
Food	2.8	49%	35
Materials handling/hydraulics	1.6	62%	42
Metal fabrication	n.a	n.a	n.a
Miscellaneous engineering	2.0	63%	52
Miscellaneous manufacturing	2.6	65%	39
Precision engineering	1.3	73%	42
Software	1.5	83%	51
Vehicles/ accessories	1.7	60%	36
Overall	2.2	63%	43

Table 14.17 shows significant differences between industrial sectors and Table 14.18 indicates that both the added value as a percentage of sales and the labour productivity is higher for the HT firms. Added value per employee averaged £27.4k for the LT group and £32.6k for the HT firms. This is quite close to the mean added value per employee of £25.9k recorded for the NI developing firms in the Performance Benchmarks (NIERC, 1996).

Table 14.18 Sales and added value (average figure) by group

	HT firms	LT firms	All firms
Turnover, £ thousands	1.5 (n=32)	2.35	2.19 (n=199)
Added value as percentage sales	76% (n=31)	62%	63% (n=185)
Added value/employee, £ thousands	32.6 (n=31)	27.4	28.9 (n= 87)

Further comparison with the NIERC Benchmarks, Table 14.19, shows that the LT companies in the sample had slightly higher added value per employee than those in the Benchmark survey but the HT firms were much more productive. It would be unjustified on this evidence alone to attribute the productivity differences solely to the difference between HT and LT firms without further investigation of other factors which may be at work. For example 90 per cent of the HT group manufactured their own products compared to 77 per cent of LT firms. A greater percentage of the LT group were sub-contractors whereas almost twice the proportion of HT were service providers, see Table 14.5. All of these factors could have an impact on opportunities to add value. Further research would be useful on these aspects.

Table 14.19 Comparison of added value per employee with benchmarks

Criteria	LT firms	HT firms	All firms
% above median Benchmark, £21,600/emp	63	75	67
% above upper quartile Benchmark, £32,600/emp	34	57	42

Universities as a Resource

There was little difference between the two groups in their attitudes to the universities as a source of assistance or advice to their businesses, 55 per cent of the LT group and 58 per cent of the HT group viewed the universities as a resource. Asked specifically about past linkages, 55 per cent claimed to have used services of a university, mainly for consultancy or problem solving. Only 2 per cent claimed to have been engaged with a university on R&D, Table 14.20.

Table 14.20 University facilities used

	HT firms		LT firms		All firms	
	Firms	%	Firms	%	Firms	%
Collaborative R&D	4	10	3	1	7	2
Consultancy	7	17	76	23	83	22
Testing	2	5	35	10	37	10
Teaching Company	2	5	11	3	13	3
Student projects	4	10	4	1	8	2
Training courses	2	5	13	4	15	4
Other	1	2	9	3	10	3

Summary

The results of an extensive programme of visits to 376 small firms in Northern Ireland, comprising 40 high-technology (HT) and 336 low-technology (LT) based firms provided data for comparison firstly between the HT and LT sectors and for later comparison with the university firms discussed in Chapter 13.

Significant differences were found between the HT and LT firms in their attitudes to competition and to markets. The HT firms being much more export orientated and much more likely to be in competition with companies outside Northern Ireland.

The HT companies were also more innovative, 61 per cent of which had introduced new products in the previous year compared with only 31 per cent of LT based firms. Expenditure on R&D was also two and a half times higher in the HT sector, although the 29 companies which supplied data spent on average only about 4 per cent of sales on R&D. It seems likely that R&D expenditure has been understated.

Product life cycle estimates showed that although, proportionately, the HT firms had twice as many embryonic and rapidly growing products, both groups had a considerable number of mature and declining products. There were however quite large sectoral variations with those firms in software, electrical engineering/electronics, agricultural engineering and food sectors having the greatest numbers of embryonic products.

Graduate and postgraduate employment was somewhat lower overall than would have been expected from the NIERC Product Development

Survey Database. Graduate employment, which was the main criteria for categorising the two groups, was 25 per cent for the HT firms compared to 3 per cent for the LT firms. The software sector was by far the largest graduate employer.

Average sales turnover in the HT group was £1.32m compared to £2.35m for the LT companies although added value, expressed as a percentage of sales and labour productivity were both higher in the HT firms. Added value per employee was also higher than that given in the Performance Benchmarks for Developing Firms (NIERC, 1996).

There was little difference between the two groups in their linkages with the universities or their attitudes to the use of university services or expertise. Fifty-five per cent of the LT and 58 per cent of HT based firms viewing the universities as a resource for their businesses, but only two per cent overall had collaborated with a university in R&D.

Note

1.   These were the same questions as were used to assess the numbers and degree of maturity of products for the university-based firms.

15 Comparison of University and Non-University Firms

Caveats

The main comparator group for the university-based firms described in Chapter 13 is the high-technology based (HT) group from the Queen's University 'outreach' exercise. It is necessary however to be cautious when comparing the results of these separate studies. Firstly, because the companies in the outreach study were all located in Northern Ireland whereas the university sample includes companies drawn from further afield. Where appropriate, therefore, comparisons have been made with the QUBIS Limited companies, which form a sub-set of about one-third of the university sample and are located in Northern Ireland.

Secondly, the age profile of the companies is different, there being a higher proportion of older firms in the outreach sample. This may influence factors such as the maturity of products and propensity to spend on research and development. Comparisons have therefore been made, in some cases, with 18 HT firms from the non-university outreach sample which are known to have been in business for less than ten years at the date of interview. Thirdly, there is a greater spread of industrial sectors amongst the university companies. The outreach HT group included a high proportion of firms in software and electronics. That however is the nature of the smaller firm sector in Northern Ireland.

Wherever possible comparisons have been made with data from other sources. This includes the study of technological firms in Europe for IRDAC, of 12 leading technology-based firms in four European countries[1] (Rothwell and Dodgson, 1987); a comparative study of firms on and off science parks (Monck et al., 1990); Performance Benchmarks for Developing Firms, (NIERC, 1996); Product Innovation and Development - comparative data from UK, Germany and Ireland[2] (Roper et al., 1996) and data from an evaluation of LEDU assistance to small firms in Northern Ireland (NIERC, 1996a). Despite the foregoing caveats and limitations of

261

the data, therefore, it is possible to draw some conclusions from these comparative results.

Competition and Markets

As expected, there were wide variations between LT and HT firms in the Northern Ireland outreach sample and many of the characteristics of the HT firms were shared by the university firms. All the HT companies recognised that they had competitors outside their region and many viewed themselves as operating in an international market. There were however some striking differences. Figure 15.1 compares the responses of the non-university LT and HT firms with those of the entire university sample and with the Northern Ireland based QUBIS Limited companies. Only 20 per cent of the latter had competitors in Northern Ireland and overwhelmingly viewed their competitors as coming from outside the region, 80 per cent noting major competitors in Great Britain and 70 per cent overseas. By contrast 54 per cent of non-university HT firms had main competitors in Northern Ireland, 54 per cent in GB and only 25 per cent overseas. (Note that Figure 15.1 shows the percentages of companies which identified major competitors in the various regions shown. The percentages do not refer to the proportion of competitors from each region.)

Similarly there was a marked difference in exports from the local region, see Figure 15.2. All of the university companies had some sales outside their local region compared with 83 per cent for the non-university HT firms. Although the non-university HT firms exported more of their products than the LT firms, they exported considerably less than the university companies. Ninety per cent of the QUBIS Limited companies, compared with only 41 per cent of the HT non-university firms, had more than four-fifths of their income from outside NI.

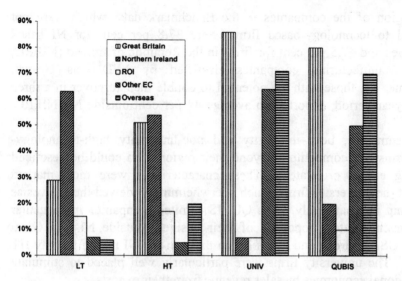

Figure 15.1 Percentage of firms identifying competitors by region

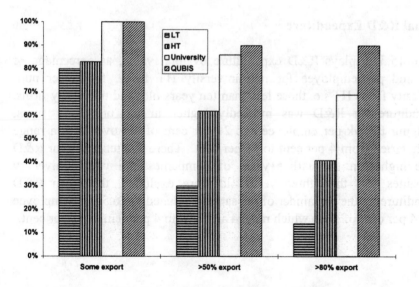

Figure 15.2 Percentage of firms with exports from their region

The average export ratio of the companies in the IRDAC study was 68.7 per cent (Rothwell and Dodgson, ibid.); while the sales destination beyond their region of the companies in the Benchmark data, which were not confined to technology-based firms, were 37.8 per cent for NI based companies and 27.5 per cent for firms in the Republic of Ireland (NIERC, 1996). Manufacturing companies classified by LEDU as Growth Companies i.e. those with the potential to double their turnover in a three to four year period, exported on average 44 per cent outside NI (NIERC, 1996a).

In summary, both university and non-university high-technology-based firms had competitors beyond their regions and could be described as being export orientated. These characteristics were most marked amongst the university firms which overwhelmingly viewed themselves as competing internationally. The QUBIS Limited companies in particular did a much greater proportion of their business outside NI - in Great Britain, USA, Europe and Japan - than those in the NI non-university HT sample. The university firms were particularly well placed to stimulate their regional economies by sales revenue from their exports.

Annual R&D Expenditure

Figure 15.3 displays R&D expenditure by total value, as percentage of sales and per employee for non-university HT firms, for newer non-university firms HT i.e. those less than ten years old, and university firms. Expenditure on R&D was markedly higher in the university firms, averaging £10.9k per employee and 24 per cent of turnover with a range which varied from 4 per cent to 95 per cent. There is a tendency for R&D to be higher in the earlier years of companies and when university companies less than three years old were excluded, the mean R&D expenditure by the remainder of the sample reduced to £6.5k per employee and 14 per cent of sales which ranged again from 4 per cent to 95 per cent.

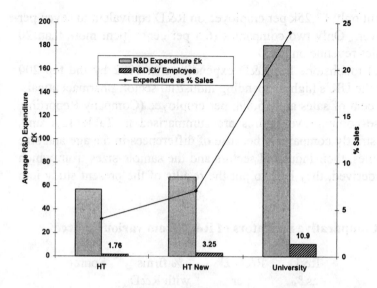

Figure 15.3 R&D expenditure by non-university HT firms and university-linked firms

Mustar (1995) noted that two-thirds of a sample of firms set up by researchers in France reported R&D expenditure of more than 20 per cent of their turnover. The figures reported in the IRDAC study showed that R&D varied from 5 per cent to 40 per cent of sales with a mean of 12.4 per cent. It was also noted that the companies with the highest proportional R&D expenditures were the younger firms in the sample. Although comparable with the figures for the older university firms in this study, the IRDAC figures are only about half of those for the university sample as a whole (Rothwell and Dodgson, ibid.).

In a survey of firms on and off Science Parks, Monck et al. (ibid.) reported that 28 per cent of on-park companies had R&D expenditure in excess of 40 per cent of sales value. This is almost exactly the figure (28.6 per cent) obtained, albeit from a much smaller sample (n=28), when the university-based firms in this study are analysed in this way. Twenty-one per cent of the university firms in the present study reported R&D expenditure exceeding 60 per cent of sales.

The HT firms in the non-university sample by contrast spent an average of £1.8k per employee and 4 per cent of sales. Even the newer non-university HT firms in the outreach sample, i.e. those less than ten years old, spent only £3.25k per employee on R&D equivalent to seven per cent of turnover. Only two companies (6.5 per cent) spent more than 20 per cent of sales revenue on R&D.

As a final touchstone the R&D expenditure in 1992 by the top 200 companies in the UK's highest spending industrial sector, pharmaceuticals, was 7.68 per cent of sales and £5.89k per employee (Company Reporting, 1993). These comparative figures are summarised in Table 15.1 and, although not strictly comparable because of differences in the age and size of the companies, their industrial sectors and the sample sizes from which the data was derived, they help to put the results of the present study into perspective.

Table 15.1 Comparative indicators of R&D from various sources

	R&D as % sales	R&D £k per employee	% firms with R&D >40% sales	Source
University firms	24.0	10.90	28.6	Chapter 13
Older univ. firms	14.0	6.50	29.4	Chapter 13
HT firms	4.0	1.80	0.0	Chapter 14
Newer HT firms	7.0	3.25	0.0	Chapter 14
IRDAC study	12.4	n.a.	n.a.	Rothwell and Dodgson (1987)
Science Park firms	n.a.	n.a.	28.0	Monck et al. (1990)
UK pharmaceutical firms, 1992	7.69	5.89	n.a.	Company Reporting (1993)

There is a much greater expenditure on R&D by these university spin-out firms than in the outreach sample of non-university HT firms and also, when the entire university sample of university firms is considered, with the IRDAC results. It is emphasised that these results are all based on relatively small samples.

New Product Introduction

Table 15.2 shows the proportions of firms in each category which had introduced at least one new product during one of the last three years[3]. Since it is to be expected that recently formed companies would have introduced new products, the table also shows data for the longer established university firms, after removing those less than three years old at the date of interview. Even when this is done, see fourth column of the table, the performance of the university firms in introducing new products exceeds that of the non-university HT sector over a three year term. The final column of Table 15.2 shows that even the newer HT non-university firms introduced a lower percentage of new products than the longer established university firms over a three year term. This cannot be unrelated to the fact that the university firms spent a much greater proportion of their sales turnover - between four and five times as much - on R&D as the non-university sample. These findings support research reported elsewhere (see Chapter 3) that high-technology firms are more innovative than others and in this comparison the university firms are most innovative of all.

Table 15.2 Percentage of firms introducing new products

New products introduced	Non-university HT firms	University firms	Older university firms	Non-university new HT firms
Last year	61%	68%	68%	64%
Or 2 years ago	24%	55%	53%	18%
Or 3 years ago	10%	48%	47%	9%

Product Maturity

Companies in the university sample were more likely to have products at an early stage in their life cycle than those in the outreach sample. There were higher proportions of university than non-university companies with products at early stages and fewer with products which were growing more slowly or declining. However when the replies from the more recently formed university companies were removed the differences were less pronounced, Figure 15.4.

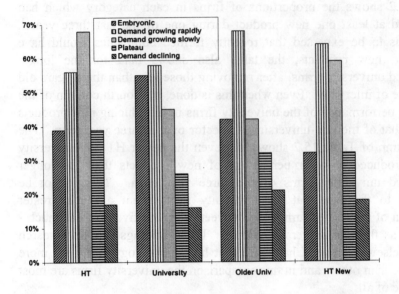

Figure 15.4 Companies having products at various stages of maturity

Table 15.3 shows this data with the percentages of companies having products at each stage as factors relative to those of the non-university HT group.

Table 15.3   Firms having products at various stages expressed as factors HT=100

Stage of maturity	Non-university HT firms	All university firms	Older university firms
Embryonic	100	141	108
Demand growing rapidly	100	104	121
Demand growing more slowly	100	66	62
Demand on a plateau	100	67	82
Demand declining	100	94	123

It is clear from the table that there is a noticeably higher proportion of university firms having embryonic and new products and rather fewer with products which have passed their best. The exception to this is the apparently low numbers of non-university companies having products for which demand was declining. Only seven companies out of 41 (17 per cent) admitted to having products in this category, almost certainly an under-estimate.

The percentage of products across all companies in both the non-university HT and university samples showed a similar trend i.e. a stark contrast with the LT firms, which had fewer embryonic and newer products but still a significant difference between the university and non-university samples even after these were adjusted for the age of the companies, see Figure 15.5.

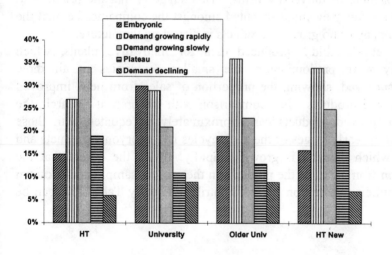

Figure 15.5 Percentages of products at various stages of maturity

Table 15.4 shows the percentage of products at various stages, by comparison with those in the non-university HT sample. Again it is clear that there are many more products at early stages in the university companies. This finding is consistent with the much greater R&D expenditure and record of new product introduction noted by the university firms.

Table 15.4   Products at various stages of maturity expressed as factors HT=100

Stage of maturity	Non-university HT firms	All university firms	Older university firms
Embryonic	100	200	127
Demand growing rapidly	100	107	133
Demand growing more slowly	100	62	68
Demand on a plateau	100	58	68
Demand declining	100	150	150

There is excellent correlation between the figures in Tables 15.3 and 15.4, the correlation coefficient is 0.93 for the older university firms and 0.95 for the entire sample of university firms. This suggests that the products at each stage are fairly evenly distributed amongst the companies i.e. that the no one company or no group of products is distorting the picture.

Roper et al. (ibid.), produced data for innovative plants, which incidentally were predominantly the smaller companies in all three countries surveyed, showing the proportion of sales from new, improved and unchanged products. For comparison with the present research, the new and improved products can approximately be equated with those products in the early stages of their life cycles i.e. embryonic products and those for which demand is growing rapidly. With the caveat that the contribution from sales of the products in the present sample is not known and the implicit assumption that all contribute equally Table 15.5 can be constructed.

Table 15.5 Comparison of life-cycle data with Roper et al. (1996)

Stage of maturity	Products of HT non-university firms	Products of all university-linked firms	Products of older university-linked firms	Percentage of sales of innovative plants
Source>	Ch. 14	Ch. 13	Ch. 13	Roper et al.
Embryonic products	15%	30%	19%	
Demand growing rapidly	27%	29%	36%	
Sub-Total	42%	59%	55%	
New products				22.8%
Improved products				29.8%
Sub-Total				52.6%

The sub-total figures for embryonic and new products and for new and improved products indicate that, with the aforementioned caveats, the university-based firms are closest in terms of their reliance on new and improved products to the innovative firms in Roper's data. The correlation coefficient between Roper's data for new, improved and unchanged products and the corresponding life-cycle stages of the university companies' products is 0.937.

STAFFING

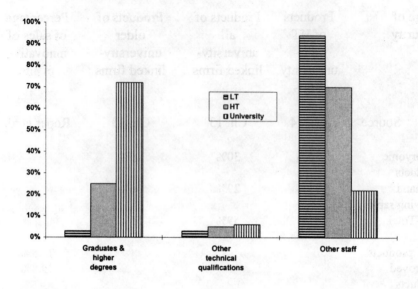

Figure 15.6 Categories of staff employed

The staffing contrasts are obvious from Figure 15.6. The high proportion of graduates in the HT sector was of course one of the ways by which that group was defined. Of more interest are the differences between the HT non-university companies and the university-based firms.

Table 15.6   Average numbers employed per firm

	Total number employed	Graduates & higher qualifications
Non-university LT firms	32	1 = 3%
Non-university HT firms	22	7 = 32%
University firms	16	11 = 69%

The higher proportions of graduates and those with higher degrees employed in the university-based firms compared to the other companies is particularly marked. On average however, the university firms employed fewer people than the non-university HT group. Table 15.6 shows the average numbers employed per firm in each category. It may be that as the university firms grow their proportions of graduates will be diluted. On the other hand, the ten largest university-based firms which employed on average 36 people, had 25 staff with first degrees or higher, equivalent to 69 per cent of their workforce. Another relevant factor may be that although the university firms concentrated almost exclusively on their core activities, many of the non-university firms were engaged in activities for which graduates and technical qualified personnel were unnecessary or inappropriate, e.g. the provision of services, sub-contracting and sales and distribution (see Table 14.5).

The proportions of graduates in all of these companies are greater than those in the NIERC Product Development Database (Roper et al. op. cit.) which showed that the percentages of employees with degrees and other technical qualifications was 20.9 per cent for UK SMEs and 18.2 per cent for Irish firms. That data was of course for SMEs of all types and not restricted to technology-based firms. A fairer comparison may be made with the science park study referred to earlier, which was confined to technology-based companies. Fifty-eight per cent of on-park firms had more than 40 per cent of qualified scientists and engineers (QSEs) compared with 36 per cent of off-park firms (see Monck et al. ibid., Table 8.3). The corresponding data from the current study show that 89 per cent of university-based firms and 26 per cent of non-university HT firms had more than 40 per cent with first or higher degrees. This suggests that the outreach HT sample of non-university companies employs fewer graduates and technically qualified people than the off-science park firms and indeed they are closer in that respect to the companies in Roper's study. The really significant differences occur in the proportions employed by the university-based firms. Further research would be necessary to test the hypothesis that although location on a science park is associated with a higher than average employment of QSEs, the participation of a university has an even greater influence on a company's propensity to employ scientists and engineers.

A total of 216 graduates plus 127 staff with higher degrees, mostly PhDs, were employed in the university-based sample. Although comparative data was not obtained for the non-university sector, it is unlikely that the proportion of those with higher degrees would be so great in the latter group.

Links with Universities

There was little difference between the LT and HT non-university groups in their attitudes to the universities, 55 per cent of the LT group and 58 per cent of the HT group claimed to regard the local universities as a resource. Despite this quite high percentage, relatively few non-university firms had actually made use of the services of the universities. The most frequently used service was consultancy, used by 22 per cent of non-university firms overall. Not surprisingly, the university spin-out firms had much closer links than the non -university groups. Links with their parent university are characterised in Table 15.7.

Table 15.7 Value of links with 'parent' university

University links	Firms	%
Close	27	87%
Remote	3	10%
Vital	8	26%
Very useful	15	48%
Useful	4	20%
Fairly useful	7	23%
Neutral	1	3%

All the university companies had frequent links with their founding institution although there was a tendency for these to become less intense as the companies grew older. Even in the oldest firms however contacts were maintained and there seemed little doubt that these ongoing contacts were mutually beneficial. In some cases these linkages gave the university companies a competitive edge through ownership or access to university-owned intellectual property. Ninety per cent of the university companies claimed to have technological 'firsts', compared to 48 per cent of the non-university HT group. Eighty-four per cent of the university firms claimed to have access to patents or unique know-how, 58 per cent of which were obtained from their parent university.

Mustar (1995) found that a group of French spin-out companies, which had been formed by researchers, maintained close links with academic institutions. These links included the publication of articles in scientific journals (more than 50 per cent), participation in scientific seminars (two-thirds of firms) and employment of doctoral students (two out of five firms).

External technical expertise and internal knowledge generation were seen as necessary complements in the overall process of technological know-how accumulation in the IRDAC study. Perhaps because of this, access to technology was not stated to be a limiting factor to growth by those longer established firms - the oldest was established in 1874 and the average age of the others was 25 years - all of which had a strong commitment to internal R&D and had established strong external links which had been progressively broadened (Rothwell and Dodgson, ibid.). Klofsten et al. (ibid.) found, for a group of Swedish spin-outs, that as time went by they became less dependent on their parent university.

The university firms in the present study were all relatively young, two-thirds being less than five years old, but it is quite likely that as they become more mature and broaden their network of external technological linkages that they will become less dependent and attached to their parent universities. Indeed the longer established firms in the sample valued their university links mainly as a means of staying in touch with academic departments to ensure a steady supply of the better graduates and appropriately qualified staff.

By recruiting graduates and postgraduates who thereby gain experiences of technology-based industry, the university companies are adding to the pool of potential future entrepreneurs. This is of some importance in a regional economy such as Northern Ireland which lacks companies from which new high-technology firms can spin-out.

Sales and Added Value

Sales turnover was almost twice as high in the non-university firms, see Table 15.8. This was largely because the non-university firms were longer established and indeed when the companies over ten years old were excluded from the non-university sector, the average turnover was reduced to £1.08m, but still 39 per cent higher than the £0.775m turnover of the spin-out companies. Bearing in mind that at least three of the university firms were little more than embryonic, this may not be such a significant difference.

Table 15.8 Turnover, material costs and added value

Average values	HT (31 firms)	HT new (16 firms)	University (29 firms)	Older university (17 firms)
Turnover, £ millions	1.5	1.08	0.775	0.761
Added value as a per cent of sales	76%	73%	82%	80%
Added value per employee, £ thousands	33	35	41	43

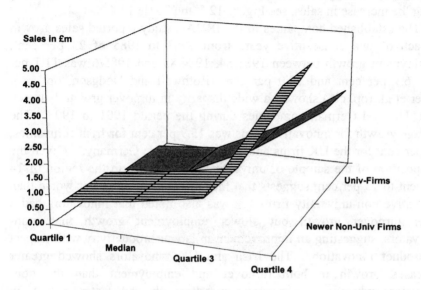

Figure 15.7 Sales by quartile for university and newer non-university firms

Figure 15.7 emphasises the different distributions of turnover between the university companies (n=27) and the newer non-university HT firms (n=16). The main differences occur due to the three largest non-university firms in the upper quartile which had turnovers of £5.5m, £5m and £2.5m respectively.

The turnover of the 18 longest established university firms i.e. those over three years old when interviewed, had grown by 15 per cent in the previous year and they were forecasting an average increase of 17 per cent for the following year. Most of the firms were interviewed at a time when the economy was just beginning to emerge from recession, a fact which may have accounted for their optimism. The longer established QUBIS Limited, companies, some of which are included in the university sample, displayed a decreasing annual rate of growth which was beginning to stabilise after the eighth year of operation at about 15 per cent annually. There was strong correlation between employment and sales growth, although some suggestion that the increase in employment was beginning to lag the increase in sales, see Figure 12.1 and Table 12.4.

The established companies in the IRDAC study reported sales growth in each of two consecutive years from 1983 to 1985 of 23 per cent. Employment growth between 1983/84, 1984/85 and 1985/6 was 11.1 per cent, 6.7 per cent and 10.0 per cent (Rothwell and Dodgson, op. cit.). Roper et al. (op. cit.) showed a wide disparity in turnover growth between Irish, UK and German companies during the period 1991 to 1993. The average growth for innovative firms was 15.9 per cent for Irish companies, 7.0 per cent for the UK firms and 1.3% for those in Germany.[4] Given the composition of the sample of university firms growth in the region of 14 per cent to 15 per cent suggests that they are neither better nor worse than innovative non-university firms. It was also noted that innovators had a faster turnover growth but slower employment growth than non-innovators, suggesting an improvement in labour productivity with respect to product innovation. The Irish and UK innovators showed greater increased growth in both turnover and employment than the non-innovators whereas in Germany innovators showed faster growth in turnover but a faster decline in employment i.e. an increase in productivity.

It is worth noting that LEDU defines growth companies as those which have the potential to double their turnover in three to four years i.e. equivalent to annual sales growth of 12 per cent to 15 per cent. All of these companies employ less than 50 people and recent research has shown that from 1989 to 1994 these LEDU assisted firms grew twice as fast as non-assisted companies in Northern Ireland and faster than surviving SMEs in Leicestershire and the Republic of Ireland (NIERC, 1996a).

The added value was about 5 per cent higher for the university firms than for the non-university HT group. The high level of added value - in the region of 80 per cent of sales - is to be expected with newer and more complex products, particularly when compared with 62 per cent for the low-technology group of non-university firms.

One of the most unexpected findings was the difference amongst the high-technology based firms is in the levels of productivity (i.e. added value per employee) which was 24 per cent higher in the university firms. Removing the younger firms, which had not yet achieved significant sales, from the university sample increased the productivity difference between the two groups to more than 30 per cent. The last column of Table 15.8

sets out the figures for the longer established university firms after removing those which had been operating for three years or less. Further research with larger samples would be required to confirm and explore the reasons for this which may be due to the higher level of human capital employed in the university firms.

Conclusions

Overview

Most of the hypotheses listed in Chapter 13, have been tested and most have held up well to the results of this investigation. The most significant differences in this comparison were to be found between the high-technology companies, whether university or non-university, and the low-technology companies in the outreach sample. These were most marked in attitudes to competition, export orientation, innovative performance, R&D expenditure, new product development and the proportions of graduate and postgraduate staff employed - which was one of the characteristics by which the firms were categorised.

Competition and Markets

The attitude to competition of many of the LT firms in the NI sample was that they deliberately positioned themselves to avoid direct competition with national and international companies (Blair and Kimber, 1995) This made sub-contracting and work with a large service element appropriate and attractive to them. Businesses adopting this strategy find exporting more difficult, R&D less important and graduate employment in technical areas of the business less vital than those with a greater awareness of the need to be internationally competitive.

In contrast, the high level of international competition and the export orientation of the university companies exceeds that of the non-university firms HT firms and also of those in the IRDAC study and the LEDU growth companies. By obtaining more of their income from outside their local regions these university spin-outs have the potential to make a much greater impact on their regional economies than even non-university high-technology based firms. Further research with larger samples would be helpful to confirm this conclusion.

Employment of Graduates and Postgraduates

There was a wide perception amongst firms which have not habitually employed graduates - found mainly amongst the LT firms but shared by a few of the non-university HT companies in the outreach sample - that graduates are ill-prepared for the world of work. This was sometimes based on little more than prejudice but it is a difficult notion to dispel. It is frequently the result of unrealistic expectations about the practical capabilities (as distinct from the knowledge) which graduates bring to the firm, but it may also be contributed to by the equally unrealistic expectations of some graduates who expect too much too soon.

Also, in smaller firms there was an observed reluctance by many owner/managers to concede that there was a need for additional management input or for succession from outside the existing ownership. This phenomenon was almost entirely confined to longer established companies in the low-technology sector. HT firms were much more willing to recruit graduates and technical staff, the university companies most of all.

One of the most striking differences between the university firms and the other high-technology-based firms in this comparison was in the number of graduates and technical staff employed. The university group employed on average almost three times as many graduates and postgraduates as the non-university HT, and it spent six times as much on R&D.

This very high level of human capital emerges as one of the most significant characteristics of university firms and it affects their performance. Although there was little difference in added value as a percentage of sales between the university and non-university HT firms (in contrast to the LT firms) the productivity of the university companies, even before adjusting for differences in the age of the two samples, was much higher. It seems reasonable to conclude that the much higher productivity of the university firms in these two samples is due mainly to the much higher level of human capital employed and further research would be beneficial to test this view.

Innovativeness and New Product Development

For many companies in the low-technology sector, innovation - in terms of R&D activity, developing and marketing new products and business methods generally, including graduate employment, where that has not been the custom - was seen as carrying unacceptable costs and associated risks. The university firms had a better record of new product introduction over a three year term than those of even the HT group of non-university firms, a greater proportion of their products were either embryonic or at the stage of rapid sales growth. They also had fewer mature and declining products.

This propensity to innovate and create new products by the university firms is not unrelated to, and may well be a direct consequence of, their proportionately higher expenditure on R&D. Roper et al. (op. cit.), showed strong correlation between R&D activity in plants and product innovation performance for companies and that innovating companies in Ireland, UK and Germany grew faster than non-innovators. Since innovation and new product development are important determinants of the competitiveness of companies one might predict that, in the longer term, the activities of university-based firms will lead to growth with beneficial economic consequences for technologically based employment.

Linkages with Universities

Links with the universities were valued and important to the university-based firms. They provide a ready source of external advice and frequently resulted in access to unique know-how and expertise. Twenty-six (84 per cent) of the university firms claimed to have access to unique patents or intellectual property and 19 of these obtained this from their founding university. Ninety per cent of these companies also claimed to have technological 'firsts' compared with 48 per cent of the non-university HT group and 30 per cent of the LT firms. It does seem therefore that continued association with a university has paid off for the spin-out companies.

Another way to interpret this data of course is that the non-university firms were more self-reliant than the university companies (Klofsten et al., 1988). Although more than half of the companies in the outreach samples claimed to have had linkages with universities, many of these contacts

282 Campus Companies - UK and Ireland

however took place a long time ago.[5] Since only 10 per cent of the HT firms claimed to have collaborated with the universities on R&D and the fact that the total R&D expenditure of this group was only 4 per cent of sales compared to 24 per cent for the university-linked firms, it appears that either the companies were under-reporting their R&D expenditure (Kleinknecht, 1987) or very little research and development was taking place.[6] It has not been possible however to test Klofsten's assertion that university companies are less robust than non-university firms and take longer to become self sufficient but as university-based firms mature it will be interesting to observe whether they follow the pattern of the Swedish firms described by Klofsten and become less dependent on their parent universities. It seems likely that in due course they will rely on the universities mainly as sources of graduates and postgraduates.

Constraints on Growth

None of the firms in either the outreach or university samples had difficulty in obtaining technical staff. Because of their close links to their parent institutions, the university companies were particularly well placed to obtain capable graduates. There appeared to be a degree of 'inside track' at work and the university firms often had privileged opportunities to attract the best graduates and postdoctoral staff. Indeed one university professor complained bitterly that the university companies were 'poaching' his research staff.

The question of funding was not addressed in during these interviews, but in the earlier series, described in Chapter 8, several universities complained about a lack of funding to pump-prime new projects. This was not widespread however and most universities felt that difficulties in establishing a market presence or in finding suitably motivated and energetic entrepreneurs, product champions or 'missionaries' were more likely than shortage of funds to constrain the rate at which their companies could grow.

Although it was widely acknowledged by the university staff in the first series of interviews that universities by themselves are generally weak at marketing, it does not follow that non-university high-technology firms are necessarily any better. Despite an admitted weakness in this area it was surprising that such a high proportion of university companies (93 per cent) attempted their own direct marketing. Further research into this apparent paradox would be worthwhile. What has been demonstrated however is that by forming equity sharing partnerships - through what QUBIS Limited refers to as corporate venturing - the ability to access markets can be greatly improved.

Growth and Mortality

It has not been possible to obtain comparative figures to assess the growth of the non-university HT outreach firms. There was fairly good agreement about turnover growth between the longer established university firms, the QUBIS Limited data, the study by Roper et al. (ibid.) and the LEDU growth companies (NIERC, 1996a). As expected, and confirming previous research (Storey et al., ibid.), most of the growth occurred in a small percentage of the firms - in the QUBIS Limited companies three firms out of 14 contributed to 84 per cent of the turnover in 1994.

None of the companies in the university-based sample is known to have failed to date although it is not known how many earlier failures may have occurred in the universities involved. Although none had failed, one of the QUBIS Limited companies stopped trading when one of the two founding entrepreneurs took up a public post with which it was felt there would be an irreconcilable conflict of interest if he continued in the university company. Statistics on mortality were not sought during the earlier interviews with universities but the general impression was that the survival rate was best in those universities which had put in place support mechanisms, such as a holding company.

Summary

The greatest differences observed were between the low-technology (LT) non-university firms and the rest. Although the university firms possessed many of the characteristics of the non-university high technology firms there were considerable differences. In almost all aspects the characteristics of high-technology firms existed to a greater extent amongst the university sample.

With caveats related to small sample size, differences in age distribution and sectoral composition of the firms, the outcome of the comparisons showed that:

- Both university and non-university firms have competitors outside of their regions but the effect was much more marked for the university firms which overwhelmingly viewed themselves as competing internationally;
- University firms were rather more export orientated than the non university firms. The QUBIS Limited companies in particular have a much greater proportion of their business outside Northern Ireland, to Great Britain, USA, Europe and Japan;
- The university firms spent a much greater proportion of their sales turnover - between four and five times as much - on R&D as the non-university sample;
- The university firms had higher proportions of embryonic or relatively new products whereas the non-university companies had more mature and declining products;
- The added value expressed as a percentage of sales was higher, at about 80 per cent in both the university and non-university HT group than in the LT group which had added value of 62 per cent of sales;
- Turnover growth of the university firms was comparable with or higher than that reported by other researchers for leading technology-based firms (Rothwell and Dodgson, 1987), innovative firms (Roper et al., 1996) and high growth firms (NIERC, 1996a);
- Productivity (i.e. added value per employee) was 30 per cent higher in the established university companies than the non-university firms;
- University firms employed a higher proportion of graduates and staff with higher degrees than other HT firms 72 per cent in this sample against 24 per cent.

Links with the universities are valued and important to university-based firms. They provide a ready source of external advice and frequently result in access to unique know-how and expertise.

Notes

1. Not all the companies in the study for IRDAC were recent start-ups, the oldest was established in 1874 and the youngest in 1981 (the study took place in 1986). Three of the firms were university spin-outs, three were mature industrial concerns, four were entrepreneurial-led independent, one was a government policy initiative and one was a joint venture. A wide range of technologies was represented. Numbers employed ranged from 24 to 1,060 and averaged 340 (Rothwell and Dodgson, 1987).

2. The report by Roper et al. (1996) outlines the major findings of a comprehensive survey of manufacturing product innovation in the UK. Also included are comparative results from similar studies in Germany and Ireland. Over 3,500 manufacturing plants contributed, of which 1,700 were located in the UK.

3. The percentages do not add to 100 because those companies which had introduced new products during the previous year did not necessarily answer about earlier years.

4. The differences in growth are mainly explained by macro-economic conditions in the three countries. The Irish GDP was growing at around 5.0 per cent, the UK was in still in recession while Germany had a downturn of 9.9 per cent in manufacturing production.

5. It is of interest that one effect of the Outreach Programme was to stimulate the use of the University's services. About 25 per cent of the visits resulted in some kind of request for assistance.

6. Some companies had difficulty estimating their R&D expenditure because it was not recorded separately in their accounts but included as part of general overhead costs. A number of firms regarded product development as an ongoing and routine activity.

16 Summary, Conclusions and Policy Implications

Summary and Conclusions

Aims and Objectives

The aims and objectives of the research, as outlined in Chapter 1, were:

- To assess the extent to which critical factors identified from the literature as influencing the formation, organisation and operation of university-based companies were relevant to current practice;
- To test the hypothesis that university-linked start-up firms share many of the characteristics of new technology-based firms (NTBFs) but that there are also some unique differences;
- To compare the characteristics of university spin-outs with non-university technology-based companies.

The specific objectives were to add to existing knowledge about university companies, their formation and management by:

- Examining the factors which influenced the universities in establishing spin-out companies;
- Exploring the motivations of actual and would-be academic entrepreneurs;
- Identifying the tasks which had to be carried out before companies were formed and how these were approached by various universities;
- Comparing the processes used and the mechanisms which various universities had put in place to carry out necessary tasks or address critical issues;
- Assessing whether best practice could be identified;
- Identifying similarities and differences between university and non-university based technology-based firms;

- Determining the extent to which technology-based university firms share characteristics with NTBFs;
- Identifying policy implications for universities, for regional development in NI and for government nationally.

Methodology

Following a review of the literature related to innovation, NTBFs and university/industry collaboration, critical factors related to the operation of university companies were identified. An extensive series of face-to-face interviews was then undertaken with universities in the UK, the Republic of Ireland, the Netherlands and Sweden to discover the relevance of these factors to current practice and to explore the experiences of practitioners who had been involved in the processes of setting up and operating university firms.

Having identified the growing importance of university-owned holding companies as vehicles for commercialising research, a number of universities adopting this approach were subsequently revisited. A detailed case-study was also carried out, of a group of spin-out firms emanating from the Queen's University of Belfast.

In order to investigate the performance and behaviour of university spin-outs, a second series of interviews was held with academic entrepreneurs and managers of a sample of university-linked firms. The results of these interviews were compared and contrasted with those from a group of companies - both high and low-technology based - located in Northern Ireland.

Main Findings - First Series of Interviews

The first series of interviews with university practitioners concentrated on the processes of company formation, the relationships between the companies and the universities and the motivations of those involved. Seven main outcomes are presented below.

Outcome 1 An enormous variety of types of university company was encountered during the interviews (Chapter 8). It is not too much of an exaggeration to say that the range, variety and types of firms established is now so diverse that many of these spin-out firms have little in common. For example, the extent of university involvement in spin-outs runs the entire gamut from relative disinterest to 'hands-on' control. It soon became apparent that, given this diversity both of types of spin-out firms and university linkages, to draw any generally applicable conclusions it would be necessary to categorise or characterise them in some way to reduce the number of variables. A method of obtaining a profile for a university/company combination, based on three significant parameters has therefore been devised and tested.

It was found (Chapter 9) that most situations could adequately be characterised by reducing the variables to three distinguishing parameters:

- The extent of university involvement in and control of the companies;
- The nature and extent of university staff involvement;
- The nature of the product or service being offered.

This observation led to the development of a new tool to enable individual university/company relationships to be 'scored' on each of these parameters. To enable this to be done consistently, degrees have been defined for each factor ranging from Degree 1 for the minimum to Degree 5 for the maximum on each factor. This has been illustrated in terms of a three-pointed star or Tripod upon which it is possible to construct a characteristic profile for a particular university/company combination, see Figure 9.1. The degree of university involvement can range from none at all to complete control. University staff involvement ranges from none to full-time commitment to the company, typically by way of secondment. The products or services produced vary from non-technological services, such as provision of university catering or conference facilities, to 'hard' companies which require a commitment to high fixed costs and a very professional approach. Between these extremes equally valid intermediate positions are possible. By joining up the scores on each arm of the Tripod a unique triangular shape can be drawn for each situation. Figure 16.1, for example, shows the typical profile of university-owned holding company holding a large but not exclusive share of company equity in a company manufacturing 'hard' products and having university staff involvement as consultants or advisers.

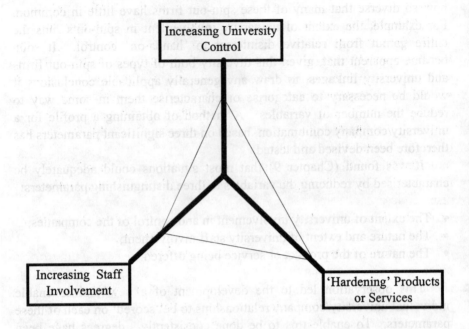

Figure 16.1 Example of characteristic profile of a university holding company

The profiling methodology, which is described fully in Chapters 9 and 10, was used to help identify the relative importance of a range of questions which must be addressed by any university contemplating the formation of new technology-based companies. A number of these questions had been raised in the literature, others came from the experiences of the practitioners as distilled from the interview responses and others suggested themselves from a priori considerations. These imply that in any university-linked company it will be necessary to have or to put in place:

- Means to assess commercial potential;
- Means to put in place entrepreneurs, product champions and managers;
- Means to decide whether and when to form companies;
- Means to interact with university departments;
- Means to interact with the wider university system;

- Means to establish Boards of Directors;
- Means to recognise and deal with conflicts of interest;
- Means to access finance and funding;
- Means to assess performance;
- Means to reward inventors/ researchers;
- Means to provide support for embryonic companies.

The relative importance of each of these questions was assessed under different combinations of the three parameters of the Tripod profile. This indicated the elements of appropriate university support structures necessary under particular sets of circumstances, see Table 10.1. Further research to promote and validate the use of the profiling methodology in a variety of different situations is proposed below.

Outcome 2 A great many universities have developed their policies and practices for new company formation - to the extent that they have thought about them at all - in a rather haphazard way and often as a pragmatic response to requests from academics to set up companies. Nevertheless, by combinations of serendipity, luck and fortuitous timing, some successful university firms have resulted from this approach. It is however rather like waiting for the muse and does nothing to ensure that potentially exploitable research is not overlooked.

Relatively few institutions were attempting systematically to exploit their research. Amongst those which were, the most successful have been those using some variant of a holding company. This is by no means the only method of managing exploitation, nor is it necessarily the most appropriate in all cases, but it is a model which is gaining in popularity and which is being successfully applied by a handful of UK universities. For those universities wishing to manage the exploitation of their R&D through new company formation, the holding company model provided the most consistent results of any encountered during this study (Chapters 11 and 12). In regions lacking a critical mass of technology based firms to sustain a regular stream of industrial spin-offs, this model is judged to be the optimum method of commercialising university research while contributing to regional economic growth.

Universities which operate a planned approach to exploitation have established infrastructures and support systems, the extent of which varies widely. This infrastructure has been described, in systemic terms, as requiring interacting sub-systems and with formalised relationships with the wider university system and the external environment. Such systems have inputs on which transformations occur to produce outputs, typically the formation of stand-alone companies.

There is however a down-side to such a prescriptive approach by the universities in that it may deter those free spirits amongst the academic community who would prefer to 'do their own thing' without university interference. In other circumstances, such as exist elsewhere in Europe and the USA where the academics own the IPR arising from their research, they are freer to take this route. Under those circumstances there is evidence that more spin-out companies will be formed but without the support mechanisms which can be put in place under more formal university structures, such as a holding company, more will fail (Chapter 12).

Outcome 3 The reasons for company formation had not always been thoroughly thought through by the institutions. Frequently companies had been formed following pressure by academics to do so and alternative means of exploiting university IPR had often not been evaluated. The opportunity costs incurred in taking the company route to commercialisation had almost never been considered.

Outcome 4 The motivations and expectations of academic entrepreneurs were frequently different from, and in some cases incompatible with, those of their institutions. This sometimes led to ambiguity or outright disagreement about how the performance of companies was to be assessed. There is a clear need for objectives to be agreed upon before company formation and, provided that the company can sustain itself from earnings or profits, there is no reason why a university-linked company should not have non-financial objectives. These however should be overt, the real danger lies in 'hidden agendas'. Most academic entrepreneurs were motivated mainly by the desire to see their research commercialised for its own sake and, although they were not disinterested in financial gain, less than one fifth rated it as the most important consideration.

Outcome 5 All the university practitioners recognised the need for and the separate roles of entrepreneurs, product champions/missionaries and managers. These roles were seldom satisfactorily performed by one person. There was also widespread support for the appointment of full-time as distinct from part-time managers, especially once the companies had progressed beyond their embryonic stage. The general experience was that academic researchers rarely made good managers.

Outcome 6 The conventional wisdom that universities are weak in marketing, although difficult to prove in objective terms, was shared by most of the university practitioners and academic entrepreneurs and was borne out in discussions. The importance of marketing was recognised also in the replies to the questionnaire on critical factors. Marketing was overwhelmingly rated as the most important factor affecting the performance of the companies. The fact that this is recognised is encouraging. One solution to this problem is the corporate venturing approach, in which the university company's equity is shared with an established firm having complementary skills to those of the university and a strong market presence (Chapters 8 and 12).

Outcome 7 Despite the emphasis on marketing, one of the distinguishing characteristics of university firms is that they more frequently begin as attempts to exploit research than as a response to a well researched market demand. Eighty per cent of the spin-outs surveyed were the result of attempts to commercialise technology and there are many examples of technology-driven firms which ultimately become extremely successful. It was notable however that many practitioners had great difficulty in distinguishing between the rather simplistic terms technology-push and market-pull. The complex and iterative nature of the innovation process is such that in practice it is not always clear at what point in a particular case a technology-driven idea begins to create or identify a market. The terms technology-push and market-pull therefore are unhelpful as other than very broad descriptors of extreme conditions of a very complex process. Usually it is only in retrospect that the route to successful innovation is clear.

A strength of this unique environment for start-ups is that such ventures are more likely to be encouraged and nurtured than they would be in the sometimes harsher commercial world. The long gestation times necessary to commercialise some types of research, such as biotechnology-based R&D, can normally only be afforded by larger and well established companies. It has been seen too that less tangible benefits are often recognised and valued by the universities, so there may be a greater tolerance of slow growth and a longer time to reach profitability than would be the case if purely return-on-investment criteria were to be applied (Chapter 8).

Main Findings - Second Series of Interviews

The second series of interviews concentrated on the performance and behaviour of a sample of university-linked firms which were then compared with a sample containing both high-technology and low-technology firms based in Northern Ireland (Chapters 13-15). The main findings are listed as outcomes 8 to 10 below.

Outcome 8 As expected, the comparison revealed enormous differences between the low-technology-based firms and all those in the high-technology sector, confirming the findings of previous research, summarised in Chapter 3. Of more interest however was confirmation of most of the hypotheses made (in Chapter 13) that university firms being a sub-set of NTBFs, would share many of their characteristics. Not only did they share most of the distinguishing characteristics of non-university technology-based firms but these characteristics were present to an enhanced degree in the university-linked firms.

Outcome 9 Amongst the most significant differences resulting from this comparison of university and non-university high-technology firms (Chapter 15) which adds to existing knowledge were:

- A much greater orientation to exports and the necessity and willingness to take on international competitors by the university firms;
- A very much greater expenditure on R&D, four times as much, by the university sample;
- Indications that the university firms were more innovative and had introduced more new products than the non-university sample;
- The extraordinary difference in employment of graduates and postgraduates in the university firms, a factor of three to one;

- Linkages with their parent universities gave the university companies privileged access to intellectual property and technological know-how;
- The productivity of the university firms was about 26 per cent higher than the non-university firms.

Outcome 10 The expectation, based on a Swedish study by Klofsten et al. (1988), that spin-outs would be less robust than non-university firms and take longer to become self-sustaining was neither proven nor disproven. Although the firms included in the present study depended less on university facilities as they matured, even the longest established companies continued to make use of university expertise and services. It may be argued that - sometimes because of the ambiguity about objectives referred to earlier (Chapter 8) - university administrations are more patient investors than those in the commercial world. Also the support offered by a holding company may shelter embryonic companies from commercial realities in a way which is not available to non-university firms.

Policy Implications for Universities

What lessons are there in all of this for the universities? Should universities become involved in new company creation and is it a good way of getting the results of research to industry? It has not been shown that new company formation is necessarily, or even usually, the most effective means either to transfer technology to industry or for a university to optimise its resources. What has been seen is that entrepreneurship is one of many ways in which this can take place and, depending on the institution's commitment to this route, it can be carried out more consistently by using a planned rather than a pragmatic approach. Whether or not a university is prepared to make that commitment will be a product of its view of how to optimise its overall resources, perceived opportunities to exploit its research and its internal ethos and public stance - for example, its commitment to assist with local economic growth.

All universities are faced with decisions about how to maximise their overall funding and resources spent on spin-out companies will not be available for other purposes, so the opportunity costs as well as the direct costs of engaging in entrepreneurial activity should be considered. By foregoing basic research, which could lead to peer-reviewed publications, a university may prejudice its HEFC research quality assessment and thereby its allocation of government core funding. On the other hand the potential income stream of profits or (more likely) the growing asset value of an investment in a successful company and the opportunity to create jobs for its graduates and postgraduates, may be regarded as the better opportunity. Decisions must be taken about the degree of commitment to this type of activity.

Universities which respond pragmatically to exploit occasional opportunities may be successful - and a few have been spectacularly so - but on balance there has been more disappointment than success. A more consistent performance has been achieved by those institutions which have adopted a systematic approach and currently a variation of the holding company model, with built-in support mechanisms, appears to offer the best way forward.

Certainly some have been successful by this means, and those which have established independent companies find that they require relatively little input either in cash or kind from the university. Once the initial investment in infrastructure has been made, all that is required is a commitment to encourage academics to bring forward their ideas and to participate either as investors, board members or technical advisers to the fledgling companies. Where necessary entrepreneurs and managers are recruited independently of the university. Queen's University's total cash investment in QUBIS Limited, for example, was less than £0.5m over a ten year term and the investment is currently valued at between £2m and £4m - a very satisfactory return.

It is more difficult to answer the second part of the question - is this an effective way to get research taken up and applied by industry? Many of the universities surveyed had not properly evaluated the alternatives and there can be little doubt that in some cases companies were formed when other methods such as collaborative R&D or licensing would have been more appropriate. More research would be needed to establish the relative strengths and weaknesses of various mechanisms of technology-transfer and in what circumstances each was most appropriate. What can be said from this research is that many new jobs have been created, an average of 15 per firm in this sample, about 70 per cent of which were for graduates or postgraduates. The companies all had a high propensity to innovate and grew at least as fast as the high growth firms with which they were compared. Where there were opportunities for the universities to license-out the technologies to local firms it is conceivable that as many jobs of comparable quality could have been created and even possibly that the technology would have been as vigorously exploited. In regional economies which lack suitable firms capable of licensing-in the new technology this could not have been done without losing it to the region. Another consideration is the importance in all of the university firms of the missionary or product champion with the energy and drive to develop the product and overcome obstacles. This personal commitment and identification with the spin-out company - which in many cases was almost tangible - would not be present in a different scenario. It is likely therefore that the exploitation would simply not have occurred by any other means. The conclusion must be that, although in some cases alternative methods of exploitation would have been equally effective, in most cases and especially in regions bereft of technology-based firms, new company formation is the most appropriate route to take.

Turning now to the processes involved, whatever type of university/company format is chosen the Tripod profiling methodology, developed in Chapter 9, has proved to be effective as a means of determining the types of support systems necessary. The relative importance of the requirements, or sub-systems, described in Chapter 10 can then be determined.

The experience of the practitioners, which is consistent with the implications of much previous research, has been that whatever type of company is envisaged there are a number of important considerations to be addressed before engaging in new company formation:

- Alternative means of exploitation should be systematically explored. This has not always been done, with the result that companies have been formed as a result of pressure from researchers without adequate consideration of the appropriateness of this route;
- The motivations and expectations of all the parties - the academics, the universities and third parties - should be explicit, understood and mutually compatible. This will lead to agreement about a new company's goals and about how performance should be measured;
- The importance of key individuals - in the roles of product champion (or missionary) and entrepreneur - was central to the success of all the companies surveyed;

- The need to employ full-time managers to run the companies and the unsuitability of most academics to fulfill this role came across strongly from the interviews.

It is clear that if universities are to become committed to the encouragement of entrepreneurial activity they will need to develop policies which take account of the aspirations of their academics. Issues such as criteria for promotion, ownership of intellectual property and opportunities to share in royalties or company profits all need to be addressed if those engaging in company related activities are not to feel disadvantaged. Procedures and systems should be put in place to lower the threshold to entrepreneurship, remove barriers and encourage researchers to come forward with their ideas. Some examples were encountered of university procedures, viewed as being too rigid or proprietorial, which drove potential entrepreneurs 'underground' and a few illicit companies have been formed by researchers who might in other circumstances have formed university-linked firms. A loose-tight structure such as that recommended by Fairtlough (1994) has much to commend it i.e. decentralised, but tight in the sense that once a consensus about priorities has been reached, everyone should respect them.

The fostering of good relationships between university firms and relevant academic departments was also of great importance in a significant number of firms. In one case there were early difficulties, interpreted as petty jealousies between those researchers involved in a company and their academic colleagues. The situation was transformed however when the company was able, because of its corporate status, to obtain government research grants which it sub-contracted to the department. In another case company staff were charged for photocopying by a hosting academic department until quid pro quo arrangements about the use of equipment were worked out. Similar anecdotes reported elsewhere emphasise the need for universities and companies to be sensitive to this type of situation.

Finally, although most of this book has concentrated on the growth of entrepreneurship, the other role of the universities in assisting companies to become and remain competitive through the diffusion of novel technology should not be lost sight of. In doing so however the universities should do more to ensure that they obtain realistic contractual terms, both for the use of their intellectual property and for research contracts. As pointed out in Chapter 5, the greatest obstacle to the universities obtaining realistic payment for their services is not the reluctance of clients to pay but the timidity of academics to ask for full costs. This is usually coupled either with a tendency for them to undervalue their own worth or to be outraged at the charges being proposed by their university administrations. This problem is widespread amongst UK universities and has frequently been debated by UDIL. Imposed price-fixing by industrial liaison staff or research offices is seldom satisfactory and certainly does nothing to change attitudes. What is required is an educational programme to point out the financial realities and dispel the widespread myth amongst academics that university overheads are somehow unreal.

Regional Implications

NTBFs are particularly important in the regional context and although much of our research was concerned with the implications of for Northern Ireland, many of the lessons learned will apply equally to other weak and peripheral regions.

Because of their potential to encourage growth of the regional economy, high-technology firms from either the private sector or universities should be encouraged and stimulated. It has been argued that, in areas lacking an adequate seedbed of existing high-technology-based firms to spawn NTBFs, university spin-outs have a disproportionately important role to play in economic development. In these circumstances the universities become the main source of new technological entrepreneurs. In NI they are by far the greatest source of new technology and innovation and consequently have an opportunity, some would say a responsibility, to become engines for economic growth.

New jobs, and particularly new technology-based jobs are badly needed in NI and the universities there are concerned to be seen to be playing a role in their creation. It has long been a problem for the province that it exports many of its most able graduates. More technology-based jobs available at home would reduce this haemorrhage. The very much higher level of human capital in the university firms is important because the graduates and postgraduates employed, most of whom are engineers or scientists, help build up the pool of potential future entrepreneurs. This will become an exponential process once a critical mass of high-technology firms becomes established. The willingness and ability to take on world-class competition and the very high proportion of goods exported from NI by the university firms also helps maximise their impact on the region's growth.

Already there have been 'second generation' spin-outs from the QUBIS Limited companies and there are signs of a clustering of indigenous software firms in NI. This however is the only discernible cluster and there may be more which government agencies such as the IDB could do to target inward investment to encourage this effect on other sectors. Knowledge-based technologies such as pharmaceuticals, biotechnology, advanced materials and medical electronics would be ideal candidates.

Even amongst existing industries there is scope for closer collaboration with universities to encourage the wider diffusion of existing technology and exploitation of new research. Food processing is one of the largest industrial sectors in the province and yet there is relatively little research done beyond the universities and the Department of Agriculture for NI (DANI) and certainly no evidence of a cluster of food technology firms such as exists, for example, around the campus of University College Cork.

Keeble (1993) has suggested the provision of government subsidised marketing and consultancy advice to stimulate regional development of NTBFs. In the NI context, in which the local universities carry out most of the R&D, this would be of less value than the provision of specific assistance to fledgling NTBFs. Financial assistance aimed specifically to assist the formation of NTBFs from both the private or university-based sector would have the potential to increase the proportion of technology-based firms in the province. Criteria for awarding assistance from this scheme should not relate to numbers employed but to an independent assessment of the technology, the commitment and energy of the missionary or entrepreneur, his/her support mechanisms and the likely market. It would however have to be administered with a light touch in such a way that decisions were given quickly and in the expectation that not all new projects would be successful. Existing support mechanisms for established firms, currently available via IRTU and LEDU would remain unaffected.

The negative attitudes to graduate employment reported by many of the owners and managers of small firms visited during the outreach programme give cause for concern (Chapter 14). There was a widespread perception that graduates lacked practical skills and that it took longer to train graduates than those who had been brought into the firm directly from school. One effect of this is that many of these SMEs were reluctant to employ recent graduates and yet they felt that they could not afford to pay the higher wages associated with experienced graduates. Graduate employment therefore was seen as an unrealistic option by many of the businesses because of the perceived high wage costs, high mobility, high training costs and higher expectations on the part of graduate employees. The fact that, in the longer run, graduates are able to contribute more by virtue of their greater knowledge and ability often was not appreciated. This is unfortunate given the high correlation observed during the outreach study between a company's innovative performance and its employment of graduates (Blair and Kimber, 1995). Similar attitudes were reported by Martin (1996) amongst some industrialists in Scotland and it would be surprising if such negative perceptions were not more widespread.

One way to tackle this would be an initiative, taken jointly by government agencies and local universities, aimed at changing attitudes to graduate employment in smaller firms, and emphasising the need for innovation to improve competitiveness. In particular, attempts should be made to dispel unrealistic expectations on the part of potential employers about the short-term practical capabilities of recent graduates and emphasise the longer term benefits. The programme could be built upon by placing a number of carefully selected graduates in 'exemplar' firms to demonstrate to others the longer term benefits which can accrue from the employment of graduates.

The DTI SMART awards have been especially beneficial in helping pump-prime university spin-outs throughout the UK and Keeble (ibid.) has shown that the take up of this scheme has been highest in the four most peripheral regions. Martin (1996) in a report to EPSRC on interactions between universities and SMEs has also drawn attention to the importance of SMART.

Local government agencies could do more to capitalise on ideas emerging from the universities. While it is understandable that a university holding company would be reluctant to take up ideas where there is no obvious missionary or entrepreneur with the energy and enthusiasm to drive the project, from a regional perspective this may mean that good ideas are being lost. More should be done to find other ways of exploiting these ideas. A regional register of technologically-oriented entrepreneurs might well be a way of matching capable people to good ideas. An example of what might be achieved is the TOS scheme at the University of Twente in the Netherlands, described by Kobus (1992), which aims to match entrepreneurs with ideas and which has directly created more than 450 new jobs since its inception in 1991. Assuming a multiplier of six new jobs in supplier companies the TOS programme is credited with the creation of 2,500 new jobs at very low cost.

Implications for Government

The effect of government grant funding on the stimulation of university spin-out firms and its relationship to other types of financial support such as that for research would benefit from further research. The tensions between the current formulaic-based research funding and commercial exploitation of research are a considerable disincentive to commercial activity by universities. Separate allocation by government of funds for research, technology-transfer and entrepreneurial activity in proportion to its perceived priorities would not be an easy exercise. Doubtless it would be regarded by many as government interference with academic freedom and would be bound to be controversial - but so too was the decision to award research funds on the basis of subjective assessment of research quality. By avoiding the issue government is deterring many universities from worthwhile commercially relevant activities in favour of conducting peer-reviewed research. There should at least be a debate about how to resolve the ambivalence of the signals coming from government. The CVCP is perhaps the most appropriate body to initiate these discussions.

There is a strong case for government development agencies to encourage high-technology entrepreneurship. The objective should be to increase the formation of new companies beyond that which would occur 'naturally'. One way would be to provide special grants for technology-based start-ups and in particular to provide seedcorn funding to enable the viability of potential projects to be tested and, where appropriate, to take projects to prototype stage so that external private funding can be obtained. Beyond that stage of development, funding to encourage R&D spending above some threshold would be helpful.

In Chapter 14 it was noted that only 2 per cent of all non-university SMEs in the NI sample reported having R&D links to the universities. This figure was only a little lower than expected and although this survey was confined to NI there is no reason to believe that it is unique to the province. It came as no surprise to the university industrial liaison staff who regarded it is indicative of the fact that there is a great reluctance amongst SMEs to engage in R&D. The experience of most universities has been that, with relatively few exceptions, smaller firms are most unlikely to undertake research other than product or process development. SMEs tend more towards the 'D' than the 'R' and many have commissioned projects in the universities which, although not categorised as R&D, have been of immense benefit to them. Despite this, governments go to great lengths to involve smaller firms in formal R&D programmes - the EU Framework Programmes in particular put great stress on the participation of smaller firms. One would have to conclude either that efforts to involve smaller companies have been singularly unsuccessful and that further research was required to improve their effectiveness or that government should have realised that the companies know best and are acting in their own best interests. The balance of evidence favours the latter view.

Further Research

Inevitably, this research has pointed to areas, enumerated below, where further research would be beneficial to build upon its significant outcomes and to explain some of its findings. The two main research-related questions emanating from the research are: the introduction of the Tripod methodology to categorise university/company relationships and the finding that the holding company is emerging as the most effective model for stimulating the formation of university-linked firms and is of particular value in assisting regional economic growth. Research topics 3 and 4, below, arise from limitations imposed on the study by the composition and geographical location of the samples of university-linked and comparator firms. Finally, research topics 5 and 6 are suggested by some unexpected findings which clearly require further study or clarification.

Research Topic 1

Further research would be necessary to test the use of the Tripod methodology, to elaborate its use under a variety of different situations and with larger samples so that a 'library' or database of relative scores and characteristic profiles could be established. It is possible too that the factor-scoring definitions included in Chapter 9 would require refinement or enlargement to deal with situations not encountered to date. With greater use and experience, and as 'typical' profiles begin to be recognised, it is believed that this technique will become of even greater value to universities and other researchers as a taxonomer of university/company groupings. It must be emphasised that scoring either high or low on any of the three parameters should not be regarded as either good or bad and says nothing, by itself, about a company's performance or prospects. The purpose of the score and the resultant profile is to help reduce the number of variables when comparing one university/company situation with another. As well as enabling support structures to be assessed, as was done in this research, the ability to compare like-with-like would be of value for inter-firm comparisons of performance between university spin-outs.

Research Topic 2

One of the most significant outcomes of this study has been that the holding company model, described in Chapter 11, is the optimum arrangement for new company formation both from the point of view of commercialising university research and of contributing to a weak regional economy. The implications of this finding are worth considering against the background of a study by Mustar (1995) who noted that, in France, the public research system is the source of a third of all new high technology-based start-ups. Seventy per cent of university spin-out firms in a French survey had achieved some financial support from local or regional government. Without such support these companies would not have been successfully established. In France, support comes from the Innovation Agency - Agence de l'innovation (ANVAR) and from a system of tax credits designed to promote research and for which newly formed technology-based firms automatically qualify. Additionally, French Government ministries responsible for research, industry and defence have opened up their research programmes and granted research contracts to one-in-three firms. Lastly, 20 per cent of the firms in Mustar's study had participated in EU technology programmes.

In the UK, the most frequently quoted and useful source of funding was the DTI SMART award scheme. Very few firms had been involved in EU-funded R&D programmes and although those in NI had taken advantage of regional grants for R&D, none of the firms encountered had received funding from the Research Councils. Although any company should avail itself of whatever government assistance is available, it is most unlikely that the degree of support available to new start-ups in France will be provided in the current competitive culture of the UK. The contradiction between the UK Government's mechanism for the funding of university research and the stimulation of university spin-out firms is another adverse effect which has already been referred to.

In these circumstances the infrastructure and support-in-kind provided by a university holding company such as QUBIS Limited is additional to, and in some cases may be acting as a surrogate for, government funding. The upsurge in the number of universities using variants of the holding company model suggests that soon sufficient data will be available for a study of the impact of holding companies on the growth rate and survival of embryonic firms. A first stage would be to determine the significant variables and to set up a series of comparisons between companies established using the holding company model and by other means. The Tripod profiling methodology would be an appropriate means to characterise the relevant variables. Being largely people-centred systems, the support structures and mechanisms provided for university companies could appropriately be studied by means of soft systems analysis (Checkland, 1981).

Research Topic 3

It has not been possible during the present research to identify the extent to which its findings are sector specific, mainly because of the diversity of technologies encountered during the interviews and in the sample of university-linked companies. Although these could possibly have been very broadly categorised into, say, engineering, software, or information technology there were too many much more significant variables which would have masked any sectoral effect. For example, many of the companies were breaking new ground with unique products in very specialised markets and only about one third had been in business for more than five years.

The much higher productivity observed in the university-linked firms i.e. 26-30 per cent higher than the non-university companies, would bear closer examination for sectoral influences. Ideally larger sample sizes and matched pairs of university and non-university firms in similar sectors should be used. The ages of the companies and their relative growth rates of turnover and employment would be significant parameters. The observation by Roper et al. (ibid.) that at some point the rate of growth in turnover outstripped the growth rate in employment amongst German companies may also be relevant, see Chapter 15.

The extent to which the higher level of human capital employed by the university firms affects their superior labour productivity should also be investigated. The stark difference between university and non-university firms in the proportions of graduates and those with higher degrees employed suggests that this may be a significant factor. It would also be of value to test the hypothesis made, in Chapter 15, that although location on a Science Park is associated with higher than average employment of QSEs (Monck et al., ibid.) the participation of a university has an even greater influence on the propensity of a technology-based firm to employ scientists and engineers.

Research Topic 4

Further examination of the impact of university spin-outs on their local economies, referred to in Chapters 3 and 15, could provide valuable insights for policy-makers in regional economic development agencies. The investigation should be extended to encompass a broad spectrum of regions ranging from disadvantaged to more prosperous areas. Ideally too the study should include international comparisons.

At one extreme the impact of spin-outs from MIT on the economy of Massachusetts was referred to by Preston (1992a) and the Cambridge phenomenon is well known Segal et al. (1990). While not discounting the importance of these findings there is an a priori case to be made that their impact could be proportionately greater in weaker economies. A possible scenario to test this would be a series of grouped comparisons between the regional role of university spin-outs firms and those of other technology-based companies - ideally matched for industry, age and size - in a number of different regions characterised by their relative economic strengths. It would be necessary to identify the possible regional benefits such as numbers and categories of staff employed, value of exports from the region, local expenditure on inputs and the use of external services supplied from within the region. The outcome of this research could well have policy implications in the context of a 'Europe of the Regions'.

Research Topic 5

Paradoxically, despite their admitted weakness in marketing, most university-linked firms concentrate on direct marketing of their products and services. The first question which could usefully be addressed is whether the widespread perception amongst university firms, in this research and elsewhere (e.g. Samsom and Gurdon, 1990; Smilor et al., 1990; Harvey, 1994), that they are less good at marketing than comparable small firms can be proved. If this really is the case the reasons should not be too difficult to find. One might speculate that the ubiquity of part-time as opposed to full-time managers may play a part and so too may the fact that these people frequently come from scientific or technological backgrounds and are untrained in marketing and business skills. Another factor may well be the absence, in the universities, of a commercial culture which recognises the prime importance of the customer.

A study of the relative advantages and disadvantages for SMEs of direct marketing, corporate venturing, selling through agencies, mail order and OEM suppliers, under various sets of conditions and for different sectors would be useful. It might help answer the question of why so many university firms appear to be so unimaginative in their choice of marketing methods. In particular, the growing practice of corporate venturing as exemplified by the QUBIS Limited companies (Chapter 12) would repay systematic study. As more university firms adopt this methodology, data will accumulate about the characteristics of successful liaisons. Topics to be investigated could include how introductions are made, the factors which make for a successful match, the benefits which each party brings to the partnership, how relationships between the parties and their relative contributions change with time and as companies mature, forms of equity sharing and necessary shareholders' agreements. The insights gained would be applicable both to university and non-university SMEs.

Research Topic 6

The final topic for additional research results from the discovery that the pros and cons of alternative methods of exploiting IPR and the circumstances in which each is most appropriate are poorly understood. Most of the UK universities were opportunistic rather than systematic in their choice of methods of transferring technology and commercialising their intellectual property. This suggests that optimum choices were not always made and the approach contrasted sharply with some US universities. MIT, for example, has very clear policies about when to license and when to form companies. Research might well indicate criteria or guidelines to help decide the methods best suited to particular circumstances. This would involve consideration of alternative policies on both the ownership of IPR and about the distribution of benefits arising from its exploitation. This research has demonstrated that there is often disagreement amongst those involved in commercial exploitation about what constitutes success. Objective measures of success would therefore have to be set from the perspectives of the potential beneficiaries, including the university, the researchers and the regional economy. Criteria could include the number and types of new jobs created - both directly and indirectly, new industries introduced in the region, export substitution, income from royalties or licence fees, profitability and return on investment.

APPENDIX

Institutions Included in Visit Programmes/Surveys

University of Aberdeen, AURIS
Queen's University Belfast, QUBIS
University of Bristol
University of Cambridge
St John's Innovation Centre, Cambridge
University College Cork (UCC)
Dublin City University (DCU)
University College Dublin (UCD)
Trinity College Dublin (TCD)
Heriot Watt Research Park
Heriot Watt University
University of Leeds, ULIS
University of Linköping, Sweden
University College London (UCL)
Imperial College London
UMIST, UVL, Manchester
University of Strathclyde
University of Surrey
University of Sussex

Other Institutions Visited

University of Pittsburgh
Massachusetts Institute of Technology (MIT)
Harvard University
National Technology Transfer Centre, Wheeling, West Virginia, USA
University of Twente
University of Lund, Sweden
British Technology Group (BTG)

Contacts and People Who Assisted

We are grateful to all those people in institutions who gave of their time to take part in interviews and otherwise to give of their expertise. Those listed below were our initial contacts but almost all of these people provided other introductions within their organisations and associated companies. Thanks are due to all of these contacts too numerous to list.

Dick van Barneveld	University of Twente, Netherlands
Edward Cartin	QUBIS Limited
Colin Dale	BTG, Edinburgh
Ian Dalton	Heriot Watt Research Park
Gillian Downie	AURIS, Univ of Aberdeen
Don Fox	AURIS, University of Aberdeen
Pat Frain	University College Dublin (UCD)
Tony Glynn	Dublin City University (DCU)
John Golds	University of Sussex
Peter Hawkes	BTG, Electronics Division, London
Walter Herriot	St John's Innovation Centre, Cambridge
Kevin Heyeck	Harvard University, USA
Carola Holmer	University of Linköping
Lani Hummel	Mid Atlantic Technology Application Centre, Pittsburgh, USA.
Richard Jennings	Cambridge University
Chuck Julian	National Technology Transfer Centre, USA
Magnus Klofsten	University of Linköping
Jan Kobus	University of Twente
Val Martinez	United States Consulate, Belfast
Lita Nelsen	Massachusetts Institute of Technology (MIT)
Eoin O'Neill	Trinity College Dublin (TCD)
Ray Oakey	Manchester Business School
Christer Olofsson	University of Linköping
Peter Raymond	UMIST Ventures Limited (UVL)
Jim Reed	University of Surrey
Grant Ross	Heriot Watt University
Clive Rowland	UMIST Ventures Limited (UVL)
Martin Sandford	BTG, London
Peter Schaefer	VUMAN (University of Manchester)

Jeff Skinner	University College London Ventures (UCLv)
Rikard Stankiewicz	University of Lund, Sweden
Peter Tanner	BTG, London
Ron Taylor	AURIS, University of Aberdeen
David Thomas	Imperial College London
Hugh Thompson	University of Strathclyde
Tony Weaver	University College Cork (UCC)
Andrew Webster	Anglia Polytechnic University, Cambridge
Ederyn Williams	ULIS Limited (University of Leeds)

Bibliography

Advisory Council for Applied Research and Development (ACARD) (1983), *Improving Research Links between Higher Education and Industry*, in collaboration with Advisory Board for the Research Councils, HMSO, London.

ACARD (1986), *Exploitable Areas of Science*, HMSO, London.

Ackoff, R.L. (1974), *Redesigning the Future*, John Wiley, London.

Advisory Board for the Research Councils (ABRC) (1987), *A Strategy for the Science Base*, a discussion document prepared for the Secretary of State for Education and Science, HMSO, London.

Academic Industry Links Organisation (AILO) (1990), *Technology Transfer - The European Experience*, proceedings of conference held at Nottingham Polytechnic 13 and 14 Sept.

AILO (1992), *Collaborative Links Between Polytechnics and Industry*, Academic Industry Links Organisation, A. Powell and T. Bomber (eds), AILO, Oxford.

American Electronics Association (1978), written statement before the House Committee on Ways and Means (E. U. W. Zachau, Chairman, Capital Formation Task Force, AEA).

Archibugi, A., Carlsson, B., Jacobson, S., Metcalfe, S. and Michie, J. (1995), *The Internationalisation of the Innovation Process and National Innovation Policies: A survey of the Literature*, first draft 5 Jan 1995, ESRC Centre for Business Research, Cambridge.

Arthur, W. B. (1990), 'Positive Feedbacks in the Economy', *Scientific American*, February, pp. 80-85.

Ashworth, J. M. (1984), *What Price the Ivory Tower: University-Industry Relationships*, The Redfearn Memorial Lecture.

AURIL (1995), *Managing Technology Transfer in UK Universities*, A. Powell and J. V. Reed (eds), AURIL, Oxford.

Bagalay, J. E. Jr. (1992), 'The University as Entrepreneur', *Industry and Higher Education*, vol. 6. no. 3, pp. 143-46.

Barden. L. (1993), 'University-Business Partnerships - Effects on Regional Economic Development', *Industry and Higher Education*, vol. 7, no. 4, pp. 200-28.

Battelle (1973), *Interactions of Science and Technology in the Innovative Process*, final report prepared for NSF, Battelle-Columbus Laboratories.

Beckers, H. L. (1992), 'Industrial R&D and Competition', the seventh Danckwerts Memorial Lecture, 2 May, *Chemical Engineering Science*, vol. 48, no. 8, pp. 1354-59.

313

Berggreen, I. (1994), 'Partnerships in Regional Development - the Bavarian Experience', *Industry and Higher Education*, vol. 8, no. 2, pp. 115-18.

Beveridge, G. S. G. (1991), 'Technology Transfer from a Regional University: Origins, Developments and Diversity', *International Journal Technology Management*, vol. 6, pp. 441-49.

Beveridge, G. S. G. and Blair, D. M. (1992), *The Challenge of Innovation*, proceedings of Technology Transfer and Implementation Conference, Day 2/ Session 5, Managing Change, 6-11, London.

Beveridge, G. S. G. and Blair, D. M. (1994), *Knowledge Transfer from a University - Its Impact on Regional Economic Development*, proceedings of Technology Transfer and Implementation Conference, Day Two, 33-39, London.

Blair, D. M. (1990a), 'Horses for Courses', *Higher Education Resources for Industry, Hobsons Directory 1989/90*, Hobsons, London.

Blair, D. M. (1990b), *Technology Transfer from a University in a Region of High Unemployment*, proceedings of International Conference on Technology Transfer and Innovation in Mixed Economies, Trondheim.

Blair, D. M. and Kimber, P. C. (1995), *Outreach Programme to Small Firms in Northern Ireland Industry*, internal report, Queen's University of Belfast.

Blair, D. M. (1996), *Aspects of Technology Transfer from Universities to Industry Through New Company Formation*, The Queen's University of Belfast, PhD Thesis.

BTG Survey by MORI (1990), into Attitudes of Academics to Commercial Exploitation of Scientific Research, The Business of Invention.

Bullock, M. (1983), *Academic Enterprise, Industrial Innovation, and the Development of High Technology Financing in the United States*, Brand Brothers and Co., London.

Burnett, J. D. (1985), *A Review of Academic/Industrial Cooperation, A Summary*, CVCP Memorandum, Cambridge University Press, Cambridge.

Butchart, R. L. (1987), 'A New UK Definition of the High Technology Industries', *Economic Trends*, February, no. 400, pp. 82-88.

Bygrave, W. D. (1995), *Can Entrepreneurship Really be Taught?*, paper presented at the Engendering Entrepreneurship in Education Conference, 26 May, Institute of Technology, Dublin.

Caborn, R. (1996), *Time to Close the Competitiveness Gap*, The Times, 12 January, London.

Cadbury, A. (1993), 'Corporate Governance', *Professional Manager*, vol. 2, no. 6, pp. 8-10, The Institute of Management.

Campbell, J. S. (1994), *Principles and Practice of Technology Foresight*, proceedings of the Innovation and Wealth Creation Conference, pp. 12-17, AILO, December, Cardiff.

Cartin, E. (1992), *Corporate Venturing - A New Tool for Technology Transfer*, proceedings of Technology Transfer and Implementation Conference, July, London.

Cartin, E. (1994), *The Growing Use of Campus Companies for Profitable Technology Transfer by UK Universities*, proceedings of Technology Transfer and Implementation Conference, July, pp. 266-271, London.

Cartin, E. (1995), *QUBIS Limited - Corporate Venturing at The Queen's University of Belfast*, unpublished paper, (available from QUBIS Ltd at Queen's University of Belfast).

Confederation of British Industry (CBI) (1970), *Industry, Science and Universities - The Docksey Report*, report of a working party on Universities and Industrial Research to the Universities and Industry Joint Committee, CBI, London.

CBI (1989), *Into the 90s - Future Success of the Northern Ireland Economy*, CBI (Northern Ireland), Belfast.

CBI (1992), *Priorities for European structural funding, 1994-98 - A strategy for Accelerated Growth to Facilitate Economic and Social Convergence within the European Community*, CBI (Northern Ireland), Belfast.

CBI (1994), *Is the UK still in the running?*, CBI News, November.

CBI (1995), *A Positive Response to the Innovation Message*, CBI News November/December.

CBI/NATWEST (1992), *Innovation Trends Survey*, CBI and Natwest Technology Unit.

Centre for Exploitation of Science and Technology (CEST) (1990a), *Attitudes to the Exploitation of Science and Technology*, London.

CEST (1990b), *Scientists' Views of the Exploitation of Science and Technology*, London.

Charles, D. Hayward, S. and Thomas, D. (1995), 'Science Parks and Regional Technology Strategies', *Industry and Higher Education*, vol. 9, no. 6, pp. 332-9.

Checkland, P. B. (1981), *Systems Thinking, Systems Practice*, John Wiley, London.

Cheese, J. (1991), *Attitudes to Innovation in Germany and Britain - a Comparison*, Centre for Exploitation of Science and Technology, London.

Council for Industry and Higher Education (CIHE) (1987a), *Towards a Partnership Higher Education-Government-Industry*, London.

CIHE (1987b), *Towards a Partnership The Company Response*, London.

CIHE (1990), *Collaboration between Business and Higher Education- Research and Development*, DTI and CIHE, HMSO, London.

CISAT (1991), *The True Price of Collaborative Research*, report of CISAT symposium, 22 January 1991.

Company Reporting Limited (1992), (1993), (1994) and (1995), *The U.K. R&D Scoreboard*, Company Reporting Limited, Edinburgh.

Constable, J. and Webster, A. (1990), 'Strategic Research Alliances and Hybrid Coalitions', *Industry and Higher Education*, December, pp. 225-30.

Cookson, C, (1989a), *Spreading the Word on Ideas*, Financial Times, London.

Cookson, C. (1989b), *Collaboration is the Name of the Game,* survey of various mechanisms including LINK and Teaching Companies, Financial Times, London.

Cooper, A. C. (1986), Entrepreneurship and High Technology, in D. Sexton and R. W. Smilor (eds), *The Art of Science and Entrepreneurship*, Cambridge, Massachusetts.

Coopers & Lybrand (1993), *Northern Ireland Economy Review and Prospects,* January, Coopers and Lybrand, Belfast.

Coopers & Lybrand (1994), *Northern Ireland Economy Review and Prospects,* January, Coopers and Lybrand, Belfast.

Corman, J., Perles, B. and Vancici, P. (1988), *Motivational Factors Influencing High Technology Entrepreneurship*, Ballinger Publishing Company, Cambridge, Massachusetts.

Cozzens, S. E. et al. (eds) (1990), 'New Roles New Linkages' in *The Research System in Transition*, Kluwer Academic Publishers, Netherlands.

Central Statistical Office (CSO) (1993a), *Business Enterprise - Research and Development, 1991*, CSO Bulletin Issue 7/93, London.

CSO (1993b), *Research and Development in UK Business*, London.

Conference of University Administrators (CUA) (1984), *Boosting University Income*, working party report.

Cullen, J. (1996), 'University-Industry Collaboration and EC Research Funding', *Industry and Higher Education*, vol. 10, no. 3, pp. 194-8.

Committee of Vice-Chancellors and Principals (CVCP) (1981), *Universities and Industry*, London.

CVCP (1988), *Costing & pricing of University Research and Contracts - The Hanham Report*, London.

CVCP (1995), *Research in Universities - Briefing Note*, London.

Danish Technological Institute (1994), *Innovative Entrepreneurs - Experiences from the Scholarship Scheme*, Taastrup.

De Benedetti, C. (1987), 'Economic Development and Industrial Structure', *International Journal of Technology Management*, vol. 2, pp 35-43.

Department of Economic Development (DED) (1987), *Building a Stronger Economy: The Pathfinder Process*, Belfast.

DED (1992), *A Research and Development Strategy for Northern Ireland - Innovation 2000*, Belfast.

De Melto, D. P. McMullen, K. E. and Wills, R. M. (1980), *Innovation and Technological Change in Five Canadian Industries*, discussion paper No 176, Economic Council of Canada, Ottawa.

Department of Finance and Personnel (DFP) (1989) *European Structural Funds - United Kingdom Regional Plans - Regional Development Plan for Northern Ireland 1989-1993*, Belfast.

De Ridder, W. J. (1995), 'Regional Policy in Technology Transfer', *Industry and Higher Education*, vol. 9, no. 5, pp. 310-12.

DFP (1993), *Northern Ireland Structural Funds Plan 1994-1999*, Belfast.

Dietrich, M. and Kurzydlowski, K. J. (1992), 'Cooperation Between Business and Higher Education in Poland', *Industry and Higher Education*, vol. 6, pp. 167-70.

Dietrich, M. and Kurzydlowski, K. J. (1993), 'A New Relationship Between Higher Education and Business in Poland', *Industry and Higher Education*, vol. 7, pp. 229-34.

Dineen, D. A. (1995), 'The Role of a University in Regional Economic Development', *University and Higher Education*, vol. 9, pp. 140-8.

Ditzel, R. G. (1990), 'Commercialising University Technology', *Les Nouvelles*, December, pp. 176-87.

Doutriaux, J. (1987), 'Growth Pattern of Academic Entrepreneurial Firms', *Journal of Business Venturing*, pp. 285-97.

Doutriaux, J. (1991), 'High-Tech Start-Ups Better Off with Government Contracts than with Subsidies: new Evidence in Canada', *IEEE Transactions on Engineering Management*, vol. 38, pp. 127-35.

Drew, R. C. B. and Edge, C. T. (1985), 'Licensing and Technology Transfer', paper read at a conference sponsored by Investors in Industry (3is), reprinted in *The Inventor*, vol. 25, no. 4 and vol. 25 no. 1.

Drucker, P. F. (1955), *The Practice of Management*, Heinemann Ltd., London.

DTI (1993), *Innovation, the Best Practice - The Report*, DTI Innovation Unit and CBI Technology Group, London.

Economist (1981), 'Getting Business and Research on the Same Track', *The Economist*, 16 May.

European Industrial Research Management Association (EIRMA) (1988), *Improving Industry-University Relations*, working group report, no. 37, Paris.

Elliott, C. (1995), 'Exploitation of Partnership? An Alternative Approach to University-Industry Collaboration', *Research Fortnight*, vol. 9, no. 1, pp.39-42.

Etzkowitz, H. (1983), 'Entrepreneurial Scientists and Entrepreneurial Universities in American Academic Science', *Minerva*, vol. 21, pp. 198-233.

Etzkowitz, H. (1990), 'The Second Academic Revolution: The Role of the Research University in Economic Development', in S. E. Cozzens et al. (eds), *The Research System In Transition*, Kluwer Academic Publishers, Netherlands, pp. 89-101.

Ewers, H. J. and Wettmann, R. W. (1980), 'Innovation-oriented Regional Policy', *Regional Studies*, vol. 14, pp. 162-79.

Fairtlough, G. (1994), 'Innovation and Organisation', in M. Dodgson and R. Rothwell (eds), *Handbook of Industrial Innovation*, Edward Elgar, London.

Fernandes, A. S. C. and Sellar, K. (1990), 'Technology Transfer from the HE Sector in Portugal', in AILO (1990), pp. 153-64.

Fielden, J. & J. M. Consultancy (1991), (unpublished), *Report to the Working Group on University-Industry Links established by Dept of Education, Northern Ireland*, John Fielden Consultancy and JM Consulting Ltd.

Financial Times (1984a), *Developing Links with Industry*, June.

Financial Times (1984b), *Bright Ideas for Inventors*, 13 July.

Financial Times (1984c), *Still too often a Dialogue of the Deaf*, David Fishlock, 5 September.

Financial Times (1984d), *Partners in Technology Transfer*, 2 October.

Financial Times (1993a), *Campus Venture Fund Unveiled*, Chris Tighe, 11 May.

Financial Times (1993b), *City University raises £20m from Technology Arm Buyout*, John Authers, 6 August.

Financial Times (1993c), *Hoping a Healthy Future will be Diagnosed*, James Buxton, 16 August.

Freeman, C. and Perez, C. (1988), 'Structural Crises of Adjustment: Business Cycles and Investment Behaviour', in G. Dosi, C. Freeman, R. Nelson, G. Silverberg and L. Soete (eds), *Technical Change and Economic Theory*, Pinter, New York.

Ganguly, P. (1985), *Small Business Statistics and International Comparisons*, Harper and Row, London.

Gemunden, H. G. and Heydebreck, P. (1994), *Technological Interweavement: A Key Success Factor for Newly Founded Technology-Based Firms*, in proceedings of TII conference Technology Transfer Practice in Europe, 18-29 April 1994, Hanover, pp. 91-106.

Gering, T. and Schmied, H. (1992), 'Technology Transfer Licensing - A comparison between Karlsruhe and France, the UK and the USA', *Industry and Higher* Education, vol. 6, no. 3, pp. 171-77.

Gibbons, M. (1984), 'Is Science Industrially Relevant? The Interaction between Science and Technology', *Science, Technology and Society Today*.

Gittleman, M. and Wolff, E. N. (1995), 'R&D Activity and Cross-Country Growth Comparisons', *Cambridge Journal of Economics*, vol. 19, no. 1, pp. 189-207.

Glynn, A. N. (1984), *Partnership in Education*, Director of Industrial Liaison, NIHE Dublin (now Dublin City University).

Goddard, J., Charles, D., Pike, A., Potts, G. and Bradley, D. (1994), *Universities and Communities - the Goddard Report*, CVCP, April, London.

Handy C. (1994), *The Empty Raincoat - Making Sense of the Future*, Hutchinson, London.

Handscombe, R. D. (1991), 'The Marketing of University Research and Development Services', *TII Focus (Newsletter of Technology Innovation Information)*, Issue 1/91.

Handscombe, R. D. (1993), 'Collaborative Research - UK Partnerships', *Industry and Higher Education*, vol. 7, no. 3, pp. 182-87.

Harris, R. I. D. (1988), 'Technological Change and Regional Development in the UK: Evidence from the SPRU database on Innovations', *Regional Studies*, vol. 22, no. 5, pp. 361-74.

Harrison, R. T. and Hart, M. (1990), 'The Nature and Extent of Innovative Activity in a Peripheral Regional Economy', *Regional Studies*, vol. 24, no. 5, pp. 383-93.

Harvey, I. (1995), 'Destroy Britain's IPR? You cannot be Serious', *Research Fortnight*, 14 June.

Harvey, K. A. (1994), *From Handicap to Nice Little Earner: A study of Academic Entrepreneurs in the UK*, paper presented at High Technology Small Firms Conference, Manchester Business School.

Hébert, R. F. and Link, A. N. (1988), *The Entrepreneur: Mainstream Views and Radical Critiques*, Praeger Publishers, New York.

Henderson, R. and Clark, K. (1990), 'Architectural Innovation: the Reconfiguration of Existing Product Technologies and the Failure of Established Firms', *Administrative Science Quarterly*, vol. 35, pp. 9-30.

Herdzina, K. and Nolte, B. (1995), 'Technological Change, Innovation Infrastructure and Technology Transfer Networks', *Industry and Higher Education*, vol. 9 no. 2, pp. 85-94.

Highfield, R. (1983), *Research Cash will Switch to Making Wealth,* Daily Telegraph.

Hitchens, D. M. W. N. and O'Farrell, P. N. (1986), *The Comparative Performance of Small Manufacturing Companies in Northern Ireland and South East England,* Department of Economics, The Queen's University of Belfast.

Hitchens, D. M. W. N., Birnie, J. E. and Wagner, K. (1992), 'Competitiveness and Regional Development: The Case of Northern Ireland', *Regional Studies*, vol. 26, no. 1, pp. 106-14.

Hitchens, D. M. W. N., Wagner, K. and Birnie, J. E. (1990), *Closing the Productivity Gap - A Comparison of Northern Ireland, the Republic of Ireland, Britain and West Germany*, Avebury, Aldershot, England.

HMG (1993), *Realising our Potential - A Strategy for Science, Engineering and Technology,* White Paper presented to Parliament by the Duchy of Lancaster. Cm 2250, HMSO, London.

HMSO (1993), *Annual Review of Government Funded Research and Development*, Cabinet Office, London.

Hobsons (1989/90), *Higher Education Resources for Industry*, Hobsons, London.

Innovation Advisory Board Action Team on Communications (IABATC) (1993), *Innovation Plans Handbook*, HMSO, London.

IACHEI (1985), *Higher Education in Support of Regional Economic and Industrial Development*, IACHEI Conference, Ennis.

Industrial Development Board for Northern Ireland (IDB) (1985), *Encouraging Enterprise - A Medium Term Strategy for 1985-1990*, Belfast.

IDB (1991), *Forward Strategy 1991-93*, Belfast.

IDB (1995), *Developing Greater Competitiveness - Industrial Development Board Strategy April 1995 - March 1998*, Belfast.

Illinois Institute of Technology Research (1969), *Technology in Retrospect and Critical Events in Science (TRACES)*, report prepared for NSF, Washington DC.

Industry and Higher Education (1995), 'Special Focus: The Role of Science Parks in University-Industry Cooperation' vol. 9, no. 6, pp. 332-57.

IRDAC (1994), *Quality and Relevance: The Challenge to European Education - Unlocking Europe's Human Potential*, Secretariat JRC, DG for Science Research and Development, CEC, Brussels.

IRTU (1993), *Industrial Research and Technology Unit : Corporate Plan 1993-1995*, IRTU, Belfast.

Irvine, J. and Martin, B. R. (1984), *Foresight in Science - Picking the Winners*, Francis Pinter, London.

Isenson, R. (1967), *Technological Forecasting Lessons from Project Hindsight*, paper read at Harvard University's Technology and Management Conference.

Isles, K. S. and Cuthbert, N. (1957), *An Economic Survey of Northern Ireland*, HMSO, London.

Kay, J. (1992), *Innovation, Technology and Competitive Strategy, Innovation Agenda - Turning Ideas into Business Success*, Number 3 in a series of 5 seminar papers on innovation, ESRC, Swindon.

Keeble, D. (1993), 'Regional Influences and Policy in New Technology-Based Firm Creation and Growth', in R. Oakey (ed.), *New Technology-Based Firms in the (1990)s*, Paul Chapman, London, pp. 204-18.

Kenward, M. (1993), 'The Fathers of Invention', *Director*, Institute of Directors, London, pp. 34-40.

Kenward, M. (1995), 'Foresight - the Saga Continues', *Director*, Institute of Directors, London, pp. 72-98.

King, B. (1990), *European Business & Innovation Centres*, proceedings of AILO Conference Technology Transfer - The European Experience, Nottingham Polytechnic, pp. 23-38.

Kinsella, R. P. (1995), *Can Entrepreneurship Really be Taught?*, paper read to Conference Engendering Entrepreneurship in Education, 26 May, Dublin Institute of Technology, Dublin.

Kinsella, R. P. and McBrierty, V. J. (1995), *Economic Rationale for an Enhanced National Science and Technology Capability*, Forfas, Dublin.

Kleinknecht, A. (1987), 'Measuring R&D in Small Firms: How Much are we Missing?', *Journal of Industrial Economics*, vol. 36, no. 2, pp. 253-6.

Kleinknecht, A. and Poot, T. P. (1992), 'Do Regions Matter for R&D?', *Regional Studies*, vol. 26, no. 3, 221-32.

Kleinknecht, A. and Bain, D. (1994), (eds), *New Concepts in Innovation Output Measurement*, St Martin's Press.

Kline, S. J. and Rosenberg, N. (1986), 'An Overview of Innovation', in *National Academy of Engineering, the Positive Sum Strategy*, National Academy Press, Washington DC.

Klofsten, M., Lindell, P., Olofsson, C. and Wahlbib, C. (1988), 'Internal and External Resources in Technology-based Spin-Offs: A Survey', in *Frontiers of Entrepreneurship Research.*

Kobus, J. A. (1992), 'Universities and the Creation of Spin-off Companies', *Industry and Higher Education*, vol. 6, no. 3, pp. 136-42.

Kuhlmann, S. et al. (1990), *The Research-Industry Interfaces in the European Community*, study undertaken for the CEC SPRINT and COMETT programmes, paper pesented at joint TII/UDIL Conference at University of Barcelona, March 1990.

Lawton Smith, H. and Atkinson, M. (1992), 'Industry-Academic Links and Local Development', *Industry and Higher Education*, vol. 6, no. 3, pp. 151-60.

Leonard, C. (1993), 'Bridging the Technology Transfer Gap', *Industry and Higher Education*, vol. 7, no. 3, pp. 176-81.

Lipsey, R. G. (1993), *Globalisation, Technological Change and Economic Growth*, Annual Sir Charles Carter Lecture, NIEC Report 103, Northern Ireland Economic Council, Belfast.

Little, A. D. (1977), *New Technology-Based Firms in the UK and the FRG*, Anglo German Foundation, London.

Louis, K. S., Blumenthal, D., Gluck, M. E. and Stoto, M. A. (1989), 'Entrepreneurs in Academe: An Exploration of Behaviours Among Life Scientists', *Administrative Science Quarterly*, vol. 34, pp. 110-31.

Lowe, J. (1986), (revised 1988), *The Management of University-based Companies and Science Parks*, report to Committee of Vice-Chancellors and Principals, CVCP, London.

Lumme, A., Kauranen, I., Autio, E. and Kaila, M. M. (1992), *New Technology-based Companies in the United Kingdom and in Finland: A Comparative Study*, SITRA, The Finnish National Fund for Research and Development, Helsinki.

Martin, B. (1995), *University/Industry Interactions*, Engineering and Physical Sciences Research Council (EPSRC).

Martin, B. (1996), *University Interactions with Small and Medium Enterprises*, Engineering and Physical Sciences Research Council (EPSRC).

Mason, C. and Harrison, R. (1994), 'The Role of Informal and Formal Sources of Venture Capital in the Financing Technology-Based SMEs in the United Kingdom', in R. Oakey (ed.), *New Technology-Based Firms in the 1990s*, Paul Chapman Publishing, London, pp. 104-24.

Metcalfe, J. S. (1995), 'Technology Systems and Technology Policy in an Evolutionary Framework', *Cambridge Journal of Economics*, vol. 19, no. 1, pp. 25-46.

McBrierty, V. J. and O'Neill, E. P. (1991a), 'The College Role in Innovation and Entrepreneurship: An Irish Experience', *International Journal Technology Management*, vol. 6, no. 5/6. pp. 557-67.

McBrierty, V. J. and O'Neill, E. P. (eds) (1991b), 'University-Industry-Government-Relations', *International Journal Technology Management*, vol. 6.

McKinsey & Company Inc. (1991), *Partners in Innovation, Business & Academia, Report for the Prince of Wales Award for Innovation*, McKinsey & Company, London.

Miner, J. B., Bracker, J. S. and Smith, N.R. (1989), 'Role of Entrepreneurial Task Motivation in the Growth of Technologically Innovative Firms', *Journal of Applied Psychology*, vol. 74, no. 4, pp. 554-60.

Monck, C. S. P., Porter, R. B., Quintas, P., Storey, D. J. and Wynarczyk, P. (1990), *Science Parks and the Growth of High Technology Firms*, Peat Marwick McLintock, Routledge, London.

Morita, A. (1992), *S does not equal T and T does not equal I*, The First United Kingdom Innovation Lecture at The Royal Society London.

Morrison, A. and Struthers, J. (1995), 'British-Russian Peer Education in Finance and Education,' *Industry and Higher Education*, vol. 9, no. 5, pp. 303-9.

Morse, R. S. (1976), *The Role of New Technical Enterprises in the US Economy*, report of the Commerce Technical Advisory Board to the Secretary of Commerce.

Mustar, P. (1995), *The Creation of Enterprises by Researchers: Conditions for Growth and the Role of Public Authorities*, OECD High-Level Workshop on SMEs: Employment, Innovation and Growth, June 1995, Washington DC.

National Academies Policy Advisory Group (NAPAG) (1995), *Intellectual Property and the Academic Community*, London.

National Board for Science and Technology of Ireland (NBST) (1980), *Innovation in Small Firms, preliminary report*, (final report published 1981), Ballymun Road, Dublin.

NBST (1986), *Barriers to Research and Consultancy in the Higher Education Sector*, Dublin.

NBST (1987a), *The Limited Liability Company as a Vehicle for Technology Transfer*, Dublin.

NBST (1987b), *Higher Education - Industry Co-operation and Technology Transfer, College Policies, Procedures and Structures*, Dublin.

NEDC (1989), *Technology Transfer Mechanisms in the U.K. and Leading Competitor Nations*, (based mainly on work at SPRU of University of Sussex).

Nelsen, L. (1994), Information provided during interview at MIT, 6 Sept 1994.

Newby, H. (1993), *Innovation and the Social Sciences Innovation Agenda - Turning Ideas into Business Success*, Number 5 in a series of 5 seminar papers on innovation, ESRC, Swindon.

Newby, H. (1995), 'Managing for Successful Innovation - The Social Science Contribution', *Industry and Higher Education*, vol. 9, no. 1, pp. 13-17.

Northern Ireland Economic Council (NIEC) (1989), *Economic Strategy; Overall Review*, Report 73, Belfast.

NIEC (1990), *Economic Assessment: April 1990*, Report 81, Belfast.

NIEC (1993), *R&D Activity in Northern Ireland*, Report 101, Belfast.

NIEC (1993a), *Northern Ireland and the Recent Recession: Cyclical Strength or Structural Weakness?*, Report 104, Belfast.

NIEC (1994), *The Implications of Peripherality for Northern Ireland*, Report 111, Belfast.

NIEC (1995), *The Economic Implications of Peace and Political Stability for Northern Ireland*, Occasional Paper 4, June, Belfast.

Northern Ireland Economic Research Centre (NIERC) (1989), *Northern Ireland Manufacturing Productivity compared with West Germany: statistical summary of the findings of a matched plant comparison*, Belfast.

NIERC (1996), *Performance Benchmarks for Developing Firms*, Belfast.

NIERC (1996a), *Evaluation of LEDU Assistance to Small Firms in Northern Ireland 1989 to 1994*, Belfast.

Northern Ireland Annual Abstract of Statistics (1994), Policy Planning and Research Unit, Department of Finance and Personnel, Stormont, Belfast.

Oakey, R. P. (1985), 'British University Science Parks and High Technology Small Firms: a Comment on the Potential for Sustained Industrial Growth', *International Small Business Journal*, vol. 4, no. 1, pp. 58-66.

Oakey, R. P. (1991a), 'High Technology Small Firms: Their Potential for Rapid Industrial Growth', *International Small Business Journal*, vol. 9, no, 4, pp. 30-42.

Oakey, R. P. (1991b), 'Government Policy towards High Technology - Small Firms beyond the Year 2000' in J. Curran and R. A. Blackburn (eds), *Paths of Enterprise - The Future of the Small Business*, Routledge, London, pp. 128-48.

Oakey, R. P. (1994), (ed.) *New Technology-Based Firms in the 1990s*, Paul Chapman Publishing Ltd, London.

Oakey, R. P. and Rothwell, R. (1986), 'High Technology Small Firms and Regional Industrial Growth' in A. Amin and J. B. Goddard (eds), *Technological Change, Industrial Restructuring and Regional Development*, Allen & Unwin, pp. 258-83.

Oakey, R., Rothwell, R. and Cooper, S. (1988), *The Management of Innovation in High-Technology Small Firms: Innovation and Regional Development in Britain and the United States*, Pinter Publishers, London.

Organisation for Economic Cooperation and Development (OECD) (1992-1995), *Main Science and Technology Indicators*, Paris.

OECD (1992a), *The Measurement of Research and Technical Activities. Proposed Standard Practice for Surveys of Technical and Experimental Development*, Frascati Manual, Paris.

OECD (1994a), *Industrial Policy in OECD Countries: Annual Review 1994*, Paris.

Olofsson, C. and Wahlbin, C. (1984), 'Technology-based Ventures from Technical Universities - a Swedish Case', *Frontiers in Entrepreneurship Research*, pp. 192,211.

Olofsson, C., Reitberger, G., Tovman, P. and Wahlbin, C. (1987), 'Technology-based New Ventures from Swedish Universities - A Survey', *Frontiers of Entrepreneurship Research*, pp. 605-16.

O'Neill, E. P. and McBrierty, V. J. (1992), 'Technology Transfer in a Changing Environment', *Industry and Higher Education*, vol. 6, no. 4, pp. 213-18.

Open University (1984), *Soft Systems Analysis: an Introductory Guide*, T301 Block IV SSA Guide, The Open University Press.

Osola, V. J. (1983). *A Critical Review of Industrial Research and Development Support Facilities in Northern Ireland The Osola Report*, J Osola and Associates, Worcester.

Office of Science and Technology (OST) (1995), *Technology Foresight - Profit Through Partnership*, London.

Oystein, F., Klofsten, M., Olofsson, C. and Wahlbin, C. (1989), *Growth, Performance and Financial Structure of New Technology-Based Firms*, paper by authors from University of Linköping for Babson Entrepreneurial Research Conference St Louis University, USA.

Packer, K. (1994), 'Patenting Activity in UK Universities', *Industry and Higher Education*, vol. 8, no. 4, pp. 243-47.

Packer, K. (1995), 'The Role of Patenting in Academic-Industry Links in the UK - Fools Gold?', *Industry and Higher Education*, vol. 9, no. 5, pp. 293-302.

Pavitt, K. (1990), 'The International Patterns and Determinants of Technological Activities', in S. E. Cozzens et al. (eds), *The Research System In Transition*, Kluwer Academic Publishers, Netherlands, pp. 89-101.

Pavitt, K. (1992), *Why British Basic Research Matters (to Britain), Innovation Agenda - Turning ideas into Business Success*, Number 2 in a series of 5 seminar papers on innovation, ESRC, Swindon.

Pereira, C. (1992), 'HE-Industry Collaboration in Northern Portugal - the Minho Experience', *Industry and Higher Education*, vol. 6, no. 4, pp. 245-49.

Philpott, T. (1994), 'Banking and New Technology-Based Small Firms: A Study of Information Exchanges in the Financing Relationship', in R. Oakey (ed.), *New Technology Based Firms in the 1990s*, Paul Chapman Publishing, London, pp. 68-80.

Pianta, M. (1995), 'Technology and Growth in OECD Countries, 1070-1990', *Cambridge Journal of Economics*, vol. 19, no. 1, pp. 175-87.

Piatier, A. (1981), *Enquette sur l'innovation: premiers resultats*, Centre d'Etude des Techniques Economiques Modernes, Paris.

Pickles, A. and O'Farrell, P. N. (1987), 'An Analysis of Entrepreneurial Behaviour from Male Work Histories', *Regional Studies*, vol. 21, no. 5, pp. 425-44.

Pike, A. and Charles, D. (1995), 'The Impact of International Collaboration on UK University-Industry Links', *Industry and Higher Education*, vol. 9, no. 5, pp. 264-76.

Porter, M. E. (1990), *The Competitive Advantage of Nations*, The Macmillan Press Limited, London.

Preston, J. T. (1992a), *Success Factors in Technology Development*, Director Technology Licensing Office, Massachusetts Institute of Technology, Revised 26 Feb 1992, basis of presentation at DTI Conference Exploiting the Science Base - The American Experience, London.

Preston, J. T. (1992b), 'Creating New Companies and Business Units within Existing Companies via University License Agreements', proceedings reported in *International Tech Transfer Business*, vol. 2.

Quintas, P. Wield, D. and Massey, D. (1992), 'Academic-Industry Links and Innovation: Questioning the Science Park Model', *Technovation*, vol. 12, no.3, pp. 161-75.

Rahm, D. (1994), 'US Universities and Technology Transfer - perspectives of Academic Administrators and Researchers', *Industry and Higher Education*, vol. 8, no. 2, pp. 72-7.

Ratnalingham, R. and Singh, C. (1993), 'HE-Industry Collaboration in Malaysia', *Industry and Higher Education*, vol. 7, no. 1, pp. 39-41.

Research Fortnight (1995), ROPA Supplement, 1, 16, 14 June.

Richards, H. (1993), *Universities create more cash than industry does*, Times Higher Education Supplement, 8 January.

Roberts, E. B. (1991), *Entrepreneurs in High Technology - Lessons from MIT and Beyond*, Oxford University Press Inc., New York.

Roberts, E. and Fusfeld, A. (1981), 'Staffing the Innovative Technology-Based Organisation', *Sloan Management Review*, Spring, pp. 19-34.

Robinson, G. (1995), 'You hear screams of agony but do you capture my soul?', *Research Fortnight*, 1 March.

Robson Rhodes (1996), *Related Companies - Recommended Practice Guidelines*, report for Higher Education Funding Councils, HEFC, London.

Roper, S., Ashcroft, B., Love, J. H., Dunlop, S., Hofman, H. and Vogler-Ludwig, K. (1996), *Product Innovation and Development in UK, German and Irish Manufacturing*, Northern Ireland Economic Research Centre, Belfast.

Rothwell, R. (1992), 'Successful Industrial Innovation: Critical Factors for the 1990s', *R&D Management*, vol. 22, no. 3, pp. 221-39.

Rothwell, R. (1993), *The Changing Nature of the Innovation Process: Implications for SMEs*, conference on New Technology-Based Firms in the 1990s, proceedings of conference at Small Firm Conference, Manchester Business School.

Rothwell, R. and Dodgson, M. (1987), *Technology-based Small Firms in Europe: The IRDAC Results and Their Public Policy Implications*, Paper at 2nd International Technical Innovation and Entrepreneurship Symposium, Birmingham.

Rothwell, R., Dodgson, M. and Lowe, S. (1989), *Technology Transfer Mechanisms in the U.K. and Leading Competitor Nations*, NEDC, London.

Rothwell, R. and Zegfeld, W. (1981), *Industrial Innovation and Public Policy: preparing for the 1980s and 1990s*, Francis Pinter, London.

Rothwell, R. and Zegfeld, W. (1982), *Innovation in Small and Medium-sized Firms*, Francis Pinter, London.

Samsom, K. J. and Gurdon, M. A. (1990). *Entrepreneurial Scientists: Organisational Performance in Scientist-started High Technology Firms*, proceedings of Babson College Entrepreneurship Research Conference, Wellesley, Massachusetts.

Scase, R. (1992), *The Innovative Organisation: Organisational Change and Competitive Advantage, Innovation Agenda - Turning Ideas into Business Success*, Number 4 in a series of 5 seminar papers on innovation, ESRC, Swindon.

Scottish Enterprise (1995), *Science and Technology - Prosperity for Scotland - Commercialisation Enquiry*, final research report for Scottish Enterprise and The Royal Society of Edinburgh, October.

Segal Quince and Wicksteed and ISI (1986), *New Technology-based Firms*, SQW, Cambridge, UK.

Segal Quince and Wicksteed, (1988), *Universities, Enterprise and Local Economic Development: an Exploration of Links*, Manpower Services Commission (MSC), HMSO, London.

Segal Quince and Wicksteed (1990), *The Cambridge Phenomenon Third Edition*, SQW, Cambridge, UK.

Sheen, M. R. (1993), 'Technology Diffusion -New mechanisms for linking supply and demand in the UK and Denmark', *Industry and Higher Education*, vol. 7, no. 2, pp. 86-92.

Skinner, G. R. B. (1989), 'Commercialisation of University-based Research - An Inventor's Perspective', *Industry and Higher Education*, vol. 3, no. 3, pp. 81-89.

Smilor, R. W., Gibson, D. V. and Dietrich, G. B. (1990), 'University Spin-Out Companies: Technology Start-Ups from UT-Austin', *Journal of Business Venturing*, vol. 5, pp. 63-76.

Smith, D. (1977), 'Contracts on the Campus', *Physics Bulletin.*

Smith, L. H. and Atkinson, M. (1994), 'The Public and Private Interface in Technology', *Industry and Higher Education*, vol. 8, no. 3, pp. 160-73.

Sonderstrom, E. J., Carpenter, W. W., and Postma, H. (1986), *Profiting From Technology Transfer: A Novel Approach*, Oak Ridge National Laboratory, USA.

Standeven, P. (1993), *Financing the Early stage Technology Firm in the 1990s: an International Perspective*, discussion paper for the Six Countries Programme Conference on Financing the Early Stage Technology Company, Montreal.

Stankiewicz, R. (1986), *Academics and Entrepreneurs, Developing University-Industry Relations*, Francis Pinter, London.

Steiner, K. V. and Kukich, D. S. (1995), 'The Academic Research Centre - a Vital Link between Industry and Higher Education', *Industry and Higher Education*,vol. 9,no. 3, pp. 135-39.

Stoneman, P. (1992), *Government Policy and Innovation Agenda - Turning Ideas into Business Success*, Number 1 in a series of 5 seminar papers on innovation, ESRC, Swindon.

Storey, D. J., Watson, R. and Wynarczyk, P. (1987), *Fast Growth Small Businesses: Case Studies of 40 Small Firms in North East England*, Department of Employment Research Paper No 67.

TCD (1995), *Innovation Services Trinity College (1995) -a general guide to the functions and activities of the Innovation Centre Trinity College Dublin*, Trinity College, Dublin.

TI (1965), *Management Philosophies and Practices of Texas Instruments*, Texas Instruments Inc., Dallas, Texas:

TII (1990), *Technology Transfer between Higher Education and Industry in Europe*, Technology Innovation Information, Luxembourg.

Times (The) (1994), *UK Science Parks*, 5 October.

Times (The) (1996), *City Technology set for Float*, Philip Pangalos, 30 May.

University Directors of Industrial Liaison (UDIL) (1988), *University Intellectual Property: Its Management and Commercial Exploitation*, London.

UDIL (1993), *Directory 1993*, London.

University of Linköping (1992), *Ideas that really mean Business*, R&D Industrial Liaison Office Universitet i Linköping , Sweden.

Vaessen, P. and Wever, E. (1993), 'Spatial Responsiveness of Small Firms', *Tijdschrift voor Economische en Sociale Geografie*, vol. 84, no. 2, pp. 119-31.

Wahlbin, C. (1989), *Growth, Performance and Financial Structure of New Technology-based Firms*, paper by authors from University of Linköping for (1989) Babson Entrepreneurial Research Conference, St Louis University, USA.

Walshok, M. L. (1994), 'Rethinking the Role of Research in Economic Development,' *Industry and Higher Education*, vol. 8, no. 1, pp. 8-18.

Walshok, M. L. (1996), 'Expanding Roles for US Research Universities in Economic Development', *Industry and Higher Education*, vol. 10, no. 3, pp. 142-50.

Weatherston, J. (1993), 'Academic Entrepreneurs', *Industry and Higher Education*, vol. 7, no. 4, pp. 235-43.

Weatherston, J. (1994), *Academic Entrepreneur*, proceedings of Technology Transfer and Implementation Conference, London, pp 296-303 (Day One).

Webster, A. J. (1988), *The Changing Structural Relationship between Public Sector Science and Commercial Enterprise*, SPSG Concept Paper no. 4, SPSG, London.

Webster, A. J. and Etzkowitz, H. (1991), *Academic-Industry Relations: The Second Academic Revolution*, a Framework Paper for proposed Research Workshop on Academic-Industry Relations, Science Policy Support Group, SGSP concept paper no.12, SPSG, London.

Westhead, P. and Storey, D. J. (1994), *An Assessment of Firms Located On and Off Science Parks in the United Kingdom*, SME Centre University of Warwick, HMSO, London.

Williamson, M. L. (1992), *Technology Transfer from Universities to Industry in the United States and the United Kingdom: Recent Initiatives and Issues for the Future*, paper read at DTI conference Exploiting the Science Base - The American Experience, 3 October, London.